29.95

PASTORAL RESOURCE COMP[illegible]

Frank E. Muhereza & Peter O. Otim

Pastoral Resource Competition in Uganda

Case Studies into Commercial Livestock Ranching and Pastoral Institutions

International Books, 2002
in association with OSSREA

Centre for Basic Research, 15 Baskerville Avenue, Kololo, P.O. Box 9863, Kampala

ISBN 90 5727 041 2

Cover design: Karel Oosting, Amsterdam
Cover photograph: M.A. Mohamed Salih, Amsterdam
Desk Top Publishing: Trees Vulto DTP en Boekproductie, Schalkwijk
Printing: Drukkerij Haasbeek, Alphen aan de Rijn

International Books
A. Numankade 17, 3572 KP Utrecht, the Netherlands
Tel. +31 30 273 18 40, fax +31 30 273 36 14, e-mail: i-books@antenna.nl

OSSREA (Organization for Social Science Research in Eastern and Southern Africa)
P.O. Box 31971, Addis Abeba, Ethiopia, phone 251-1-551 163
e-mail: ossrea@telecom.net.et, http://www.ossrea.org

Table of Contents

List of Maps

List of Tables

Acknowledgements

These studies were made possible as a result of a grant provided by the Organisation of Social Science Research in Eastern Africa (OSSREA) Addis Ababa, in conjunction with the Institute of Social Studies (ISS), at the Hague, through funding from the Research section of the Directorate General of International Co-operation of the Netherlands Ministry of Foreign Affairs (DGIS), for which we are most grateful.

At the beginning of this study, three volumes of carefully selected secondary reading material on pastoralism and resource competition which are available in the Library at the Institute of Social Studies at the Hague, were compiled by Professor Mohamed Salih and sent to each of us on the study team. The first volume covered world material on the subject, the second on Africa and the third on Uganda. We are grateful for these resources.

We are also grateful to the ISS and OSSREA for the hospitality accorded to us while on a two-month fellowship at the ISS between June and August, 1998. (Especial thanks go to Ms. Els Mulder and John Sirjonge for facilitating our stay at The Hague) and for the Methodological Workshop held in 1996 in Addis Ababa, respectively.

To the various people without whose help access to certain information would not have been possible, especially in the Ministry of Agriculture, Animal Industry and Fisheries, the district headquarters in Mbarara and Moroto, we are most obliged.

Lastly, we wish to thank the various people we interacted with in the course of collecting data for writing this manuscript, including those who agreed to be our guides and respondents in Kampala, Mbarara, Masaka, Luwero, Nakasongola, Masindi and Moroto districts.

About the Authors

Frank Emmanuel Muhereza is currently a Research Fellow at the Centre for Basic Research Kampala. Muhereza holds a first degree in Political Science and Public Administration from Makerere University obtained in 1990, and a Diploma in Development Studies from Cambridge University obtained in 1993. He has just completed a Masters of Science in Environment and Natural Resources from Makerere University, Kampala.

Peter Omurangi Otim is also currently a Research Fellow at the Centre for Basic Research Kampala. He holds a first degree in Sociology from Makerere University obtained in 1990, and an MPhil in Social Anthropology obtained from the University of Bergen in Norway in 2000.

Abbreviations and Synonyms

ADA	Assistant District Administrator
ADB	African Development Bank
ADC	Aide-de-camp
APC	Armoured Personnel Carriers
ASTU	Anti Stock Theft Unit
CBPP	Contagious Bovine Pleural-pneumonia
CGR	Central Government Representative
DC	District Commissioner
DVO	District Veterinary Officer
EAP	East African Protectorate
ECF	East Coast Fever
FAO	Food and Agricultural Organisation
IPDP	Integrated Pastoralist Development Programme
IIED	International Institute for Environment and Development
ISS	Institute of Social Studies
KDA	Karamoja Development Agency
KPIU	Karamoja Projects Implementation Unit
LDF	Local Defence Forces
LMNP	Lake Mburo National Park
LSP	Livestock Services Project
LUIU	Land Use Investigation Unit
MAAIF	Ministry of Agriculture, Animal Industry and Fisheries
MAIGF	Ministry of Animal Industry, Game and Fisheries
MALIFA	Masaka Livestock Farmers Association
NEC	National Executive Committee
NGO	Non-Governmental Organisations
NRA	National Resistance Army
NRC	National Resistance Council
NRM	National Resistance Movement
NWC	National Wildlife Committee
ODI	Overseas Development Institute

OSSREA	Organisation of Social Science Research in Eastern and Southern Africa
RAC	Ranches Advisory Committee
RDC	Resident District Commissioner
RPAB	Ranches Policy Advisory Board
RRB	Ranches Restructuring Board
RSB	Ranches Selection Board
RVO	Regional Veterinary Officer
UHT	Ultra Heat Treated (milk)
ULC	Uganda Land Commission
UNDP	United Nations Development Programme
UNRISD	United Nations Research Institute for Social Development
UNSO	United Nations Sudano-Sahelian Office
UP	Uganda Protectorate
UPC	Uganda People's Congress
UPSMP	Uganda Protectorate Secretariat Minute Paper
USAID	United States Agency for International Development
VSAI	Veterinary Services and Animal Industry, (Department of)
WUC	Water Users' Committee

CHAPTER ONE

Introduction

1.1 The Case Studies

Within concrete socio-cultural, physical and ecological conditions, interventions not only through policies but also politics, have variously influenced the practice of cattle keeping in different communities. Livestock production policies have affected the manner in which cattle keepers organise their production systems as much as the political factors, in the extreme, have shaped the outcome of pastoral policies and the various ways in which pastoral resources are utilised.

Cattle keepers in different socio-cultural, physical and ecological conditions have been the subject of control by the state. This control has been exercised mainly through policies designed to transform pastoral production. While in Ankole interventions were justified mainly because of ecological considerations concerning rangeland degradation and the improvement of the efficiency of the utilisation of rangeland resources, in the broader Karamoja region such intervention were intended to 'pacify' an otherwise war-like population of moribund cattle keepers. Specific policies intervened in the livestock sub-sector in different production systems mainly because different governments in power identified different problems and solutions to livestock production. However, the underlying objectives have largely remained unchanged.

These two case studies of Ankole and Karamoja are intended to create a new understanding of how prevailing pastoral systems have been shaped, not only by the physical/ecological conditions, but also - and more importantly – by human interventions. The Ankole region comprises vast cattle keeping corridors located within the country's dry savannah belt in the southwestern part of Uganda, stretching from Uganda's southern border with Tanzania up northwards to Mbarara district. This region comprises the country's leading livestock production area. Karamoja region, on the one hand, comprises the districts of Moroto and Kotido, located in the northeastern part of the country. Livestock production in the two livestock producing regions provides a very interesting contrast to how policies can lead to different outcomes in dif-

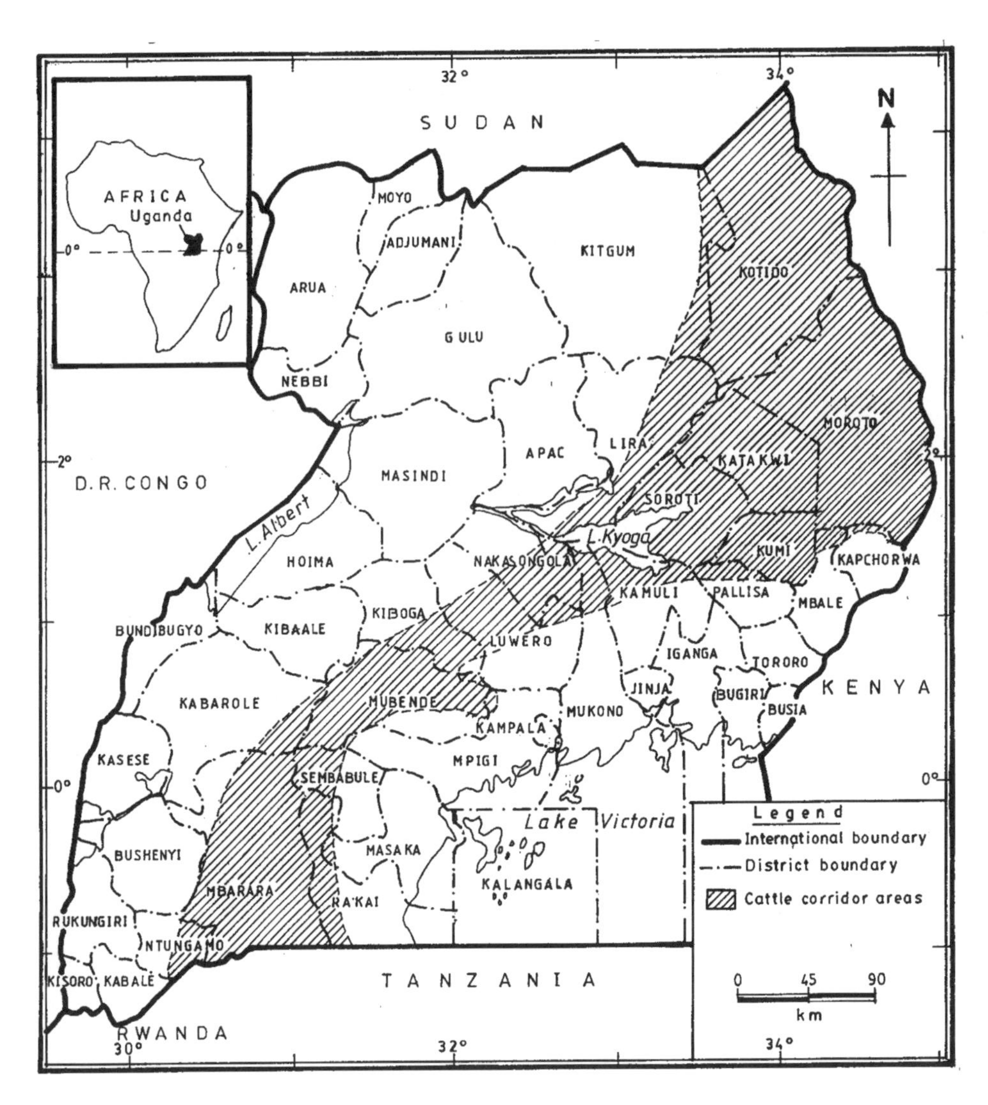

FIGURE 1 *Map of Uganda Showing the Case Study Areas*

ferent environments, with wider implications to the respective societies. While policies and politics achieved different outcomes in the case of the two societies of Ankole and Karamoja, both were intended to influence the practice of pastoralism in order to transform pastoral production from what was considered 'traditionalism' in pastoralism to 'modernity'.

The obsession of policy makers with the cultivation culture – and therefore sedentarisation – as the best mode of land use has largely been responsible for this. The pastoral way of life has not been looked at positively by policy makers. As it was not the mainstay of the national economy, pastoralism has been regarded as a secondary economic activity. The lifestyle of cattle keepers has in particular been considered incompatible with civilised ways of life (UNSO/UNDP, 1993). Their mobile lifestyle is seen as backward and destructive, and so most government policies are aimed at changing the pastoral way of life – using the models considered appropriate by outsiders.

In the Ankole case study, Frank Muhereza has argued that since the colonial period the overriding concern of the state has been the commercialisation of livestock production through the introduction of commercial livestock ranching. Different governments pursued different forms of intervention intended to achieve commercial livestock ranching. According to Muhereza this implies that it is significant to understand the specific goals and objectives for which the various governments sought to develop the livestock sector.

Muhereza argues that the role played by the state is important in understanding the rationale for policy interventions. To a great extent, the implications these policies have on the allocation and use of available scarce resources, the possible responses by those affected by the respective policies also need to be examined. Sometimes even well- intentioned policy interventions in the livestock sector have resulted in undesirable consequences both to the cattle keepers and the range-lands (Raikes, 1981). This suggests that both the overt and sometimes the covert objectives and intentions of the policies for the development of the livestock sector need to be investigated.

In the case study on Karamoja, Peter Otim shows how the current institutions in pastoral management in Karamoja function, how they have developed over the years, and how they are utilised. Otim provides insights on how such institutions have evolved over the years, drawing mainly on the impact that the proliferation of firearms has had on the changing nature of organisation of herd management systems. The analysis of the changing nature of the *Adakarin* (the herding co-operative) is central to the argument made by Otim in this case study. Otim underlines the internal military organisation of the *Adakarin* and what it has intended to achieve, as well as how this herding institution has fostered the adaptability of the Ngakarimojong people.

Otim's argument draws on the human adaptability theses. He shows how humans try to adapt to factors external to their social, political and physical environment that impinge on their options for survival. Otim also shows external factors cause changes in human institutions. People try to adapt to these changing circumstances as well. In Karamoja, Otim argues, the organisation of cattle keeping has also been greatly influenced by regional political factors. Especially insecurity in the region caused by a combination of factors such as the war in Southern Sudan and the competition for scarce grazing resources with other armed cattle keeping groups from Kenya and Sudan has greatly influenced the organisation of cattle keeping.

It has been argued in the case study on pastoral institutions in Karamoja, that communities/societies develop institutions that are meant to foster their livelihoods through adaptive survival strategies. Changes occasioned in these institutions, resulting from intervention external to the pastoral systems, have many times affected the institutions negatively, especially if they constrain the smooth functioning of the adaptive survival strategies.

These case studies contribute to the discussions of the problems of policy interventions in the development of the livestock sector. The fact that these policy interventions contribute to changes in resource availability between different stakeholders implies that they cannot be considered 'neutral'. As far as these studies are concerned, the reasons that have been given for certain policy failures are not 'neutral' either, and should therefore be unpacked. Livestock sector development policies, how they are implemented, how those, affected by these policies, respond to them, and how the problems and failures of these policies are explained, have implications that are political in nature. In addition, the implications of these policy are influenced by general political economic tendencies at a specific point in time.

This manuscript is divided into four chapters. Chapter one is the introduction, which sets the context within which the two case studies on livestock production in Uganda can be understood. Chapter one provides an overview of livestock production in Uganda, showing land utilisation for livestock production. A discussion of the potential for livestock production in the different agro-ecological zones is presented as well.

Chapter two focuses on the Ankole case study discussing the history of commercial livestock ranching in Uganda since the colonial period, and the implications this has had on the type of livestock production in the region. The chapter examines the genesis of commercial livestock production in Uganda, showing how the colonial state justified the need to develop the sector. It also shows how the subsequent privatisation of tenure in the rangelands was justified by the colonial administration. It analyses the development of livestock policies in the period after independence. It discusses the manner in which the Obote I government implemented colonial policies of establishing government commercial livestock ranching schemes. The controversies surrounding the actual implementation of the policy are discussed, especially the problems of allocating the ranches. This chapter also discusses how the development of the ranches progressed, the implications of the terms and conditions imposed on ranch development, the resistance to the establishment of the ranching schemes, and the achievements that were registered by the ranching schemes.

Chapter two also looks at the development of commercial livestock production between 1972 and 1979. This period not only saw the sector attain peak cattle population numbers, but also the beginning of the downward trend, largely caused by political and economic problems. The attempts to revamp the livestock production sector during the Obote II government and the limitations that were faced are also discussed. This section also discusses the cause, magnitude and implication of the problem of squatter pastoralists on the ranching schemes. The nature of livestock development during the NRM administration starting in 1986, including the de-gazetting of part of the Lake Mburo National Park to settle landless cattle keepers is discussed. The repossession of former government ranching schemes, and their subsequent restructuring to make land available for settlement of landless cattle keepers (who were still landless) is also undertaken. The study specifically underlines the various arguments used by the government to justify the various policy interventions. In this section the study also shows how these policies were played out in the allocation of land made available from ranch restructuring. In section seven the study examines the current prospects and future constraints to livestock development in Uganda, in the light of some of the livestock development policies undertaken by the NRM government. The study specifically looks at water development and the increasing adoption of settled crop cultivation.

Chapter three focuses on the case study on Karamoja. Chapter three contains a descriptive analysis of key local pastoral institutions of the Karimojong. It looks at how they have fostered their adaptation to the semi-arid environment amid various constraints of the natural environment, of politics, and of economics both from the local and larger external levels they interact with. Decision-making is a critical factor as individuals pursue those objectives that enable them to survive. In this chapter, the central focus is on the traditional pastoral institutions such as the age system. It shows how shifts and changes in political authority have an impact on the decisions and options that are critical for the survival of individuals. An attempt is made in this chapter to show the influence of local and external factors in the shifts and changes of power and authority and its impact on the adaptive process. Various factors external to the Karimojong habitat have influenced this traditional leadership. These include the state (right from the colonial era) and the proliferation of arms resulting from political instability in the Eastern Africa region in general. The analysis is drawn from ethnographic field research carried out in the village of Rupa in Rupa sub-county and Matheniko County in the district of Moroto.

The concluding remarks are made in Chapter four. This chapter contains a summary of some of the major arguments made in the two case studies.

1.2 An Overview of Livestock Production in Uganda

1.2.1 Land Utilisation for Livestock Production

Uganda is located in the eastern part of Africa astride the eastern and western arms of the East African Rift valley systems and lies between latitudes 4°7° north and 1°3° south, and between longitudes 29°33° west and 35°20° east. The total area of Uganda is about 24.1 million hectares, of which 19.4 million hectares (82 percent) is land area. Of this land area, arable and permanent crops occupy about 5.5 million hectares, pastures and grazing land 5.0 million hectares, and forest 6.5 million hectares, of which 1.5 million hectares has been reserved forests. The balance of 2.4 million hectares is comprised of rocky mountain slopes, Game Reserves and National Parks.

Livestock production is a crucial part of Uganda's agriculture and accounts for a 17 percent contribution of the Gross Domestic Product (Republic of Uganda, 1997). This contribution is about 9 percent of the Gross Domestic Product (GDP), although performance has been declining over the years. Eighty percent of the herds of cattle is currently found in southern and western Uganda, where the average number of cattle per household is 2.11 compared to northern Uganda where it is 0.67, with the national average at 1.37 (Fintecs Consultants, 1997).

Of all the livestock, cattle are the most important, although goats, and to a lesser extent sheep, pigs and poultry make significant contributions to the local economy and diet (Republic of Uganda, 1997). It is on livestock that this study focuses, although references in the text of this study will often be made to livestock production. Cattle are reared mainly in the range-lands which occupy about 84,000 square kilometres extending from the more arid Moroto and Kotido in the north-east through the sub-humid flats of lake Kioga to Masaka and Ankole. These areas experience unreliable rainfall with a long dry spell from October to March. The mean annual rainfall varies from 500 mm to about 1500 mm with high degrees of fluctuation between the years and locations. Average temperatures range from 18 °C to 20 °C, with a maximum of 28 °C and 30 °C. The majority of these savannah areas are marginal for crop production and have a fragile ecology where the grass grows very fast during the rainy season loosing its nutritive value very quickly. These areas are sparsely populated and mainly inhabited by nomadic cattle keepers (Fintecs Consultants, 1997).

The cattle reared include indigenous and improved (crosses) as well as exotic breeds. The indigenous breeds are mainly kept under the extensive traditional production system. Breeds include the east African short horn Zebu, the longhorn Sanga-Ankole, and the intermediate Ganda cattle. Other indigenous breeds have originated from Kenya (Boran and Turkana) and Sudan (Toposa). Exotic beef breeds originating from the United Kingdom, Denmark and the Netherlands include Charollais, Hereford, Aberdeen Angus, Sussex, and Shorthorn. The main dairy breeds kept are Friesians. Improved breeds are mainly crosses between indigenous and exotic breeds. They are kept under intensive management, mainly on small scale and medium sized farms and zero grazing.

Karamoja strikes one as a flat expanse of territory with dotted thorn scrub – generally scarce, short vegetation bearing evidence of heavy grazing – quite in sharp contrast with its neighbouring regions. On entering Karamoja, one can also see a few scantily dressed individuals walking along the roads, the males in most cases carrying AK-47 assault rifles on their shoulders apparently oblivious of what is going on around them. Occasionally one sees mainly young boys with a few cattle grazing by the roadsides, but in general, this region is so quiet that it makes one wonder where the livestock of these pastoralists are.

This region normally referred to as Karamoja covers the two districts of Kotido and Moroto in the north eastern part of Uganda (see map). This semi-arid territory is about 300 kilometres in length from north to south with a narrow width averaging about 80 kilometres from east to west. It is a relatively flat plain punctuated by some hills and mountains that include Mount Moroto in the east reaching up to 10,114 ft. To the west are the Akisim and Napak mountains overlooking Katakwi district. Toward the southern border is Mount Kadam.

The main ecological limitation in this region is its inadequate and unreliable rainfall. It is not only little – averaging 350 mm to 750 mm per annum, but also unreliable with regard to when it falls, how much and for how long it falls, and which area it will cover. What comes out as the determining factor for the

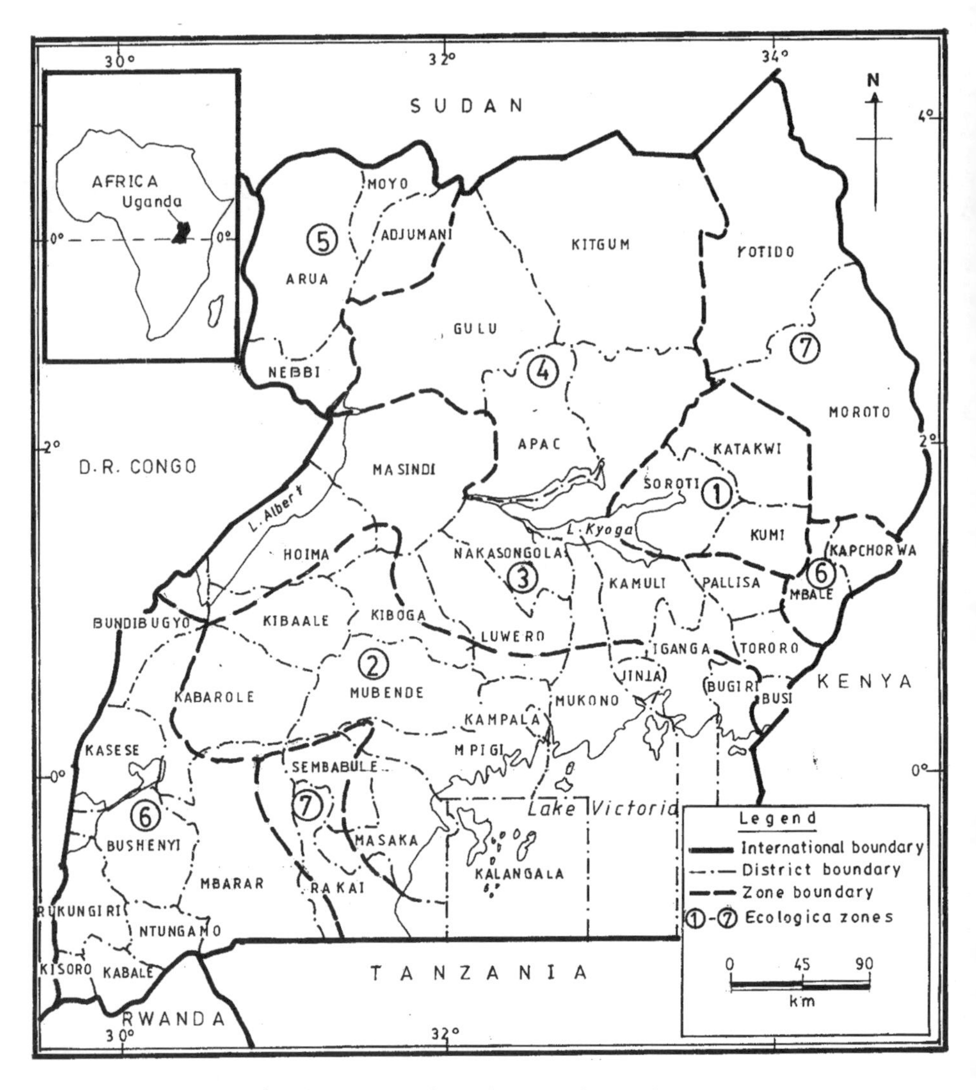

FIGURE 2 *Map showing Agro-ecological Zones of Uganda*
Source: Meat Production Master Plan, 1997

1 = Teso; 2 = Banana/Coffee; 3 = Banana/Cotton; 4 = Northern; 5 = West Nile; 6 = Montane; 7 = Pastoral

ecological features of Karamoja is this unreliable rainfall. The rains are scattered, varying in amount from year to year and even from one place to another in the same year (Dyson-Hudson, 1966; Mamdani, *et al*, 1992; Cisteriono, 1979; Welch, 1969). While one place receives a sprinkle, another receives a heavy storm. The result of this pattern of rainfall is a low resource base characterised by seasonal variations in productivity presenting patchy conditions even within the same zone; there is luxuriant vegetation in one location and near emptiness in another.[1]

Karamoja has numerous rivers that appear as sandy depressions during the dry season and occasionally become fast flowing rivers during the rainy season. The rain falls in torrents and this causes the rivers to swell within a few hours and roar downstream – sometimes through dry areas, sweeping away people, livestock, and whatever else crosses their way. This usually lasts from a few hours to a few days depending on the amount of rainfall. Then the riverbed dries up again leaving only a few ponds in view. Generally the region slopes westward so that most of the rivers flow to the west eventually feeding the perennial swamps that form the boundaries with Katakwi, Kumi, Lira, and Kitgum. The naturally occurring water catchment areas are small in size and hold water only for a short time; the high temperatures in the region encourage quick evaporation.

Such climatic conditions as described above have not favoured crop cultivation. Whereas the pastoralists in Karamoja are known to have practised agriculture for a long time (Gulliver, 1955; Cisterino, *op cit.*; Lamphear, *op cit.*; Ocan, *op cit.*; Muhereza, 1995.), it has been characterised by crop failure resulting from the unreliability of the rainfall. Under such conditions, the only option for viable agriculture in this region would be irrigation, but because irrigation is not practised, the people depend on rain-fed agriculture. As a result agriculture is not viable. This has limited the options available for survival in the region. In this environment where four out of every five crops fail (Mamdani et al, *op cit.*; Dyson-Hudson, *op cit.*; Cisterino, *op cit.*), pastoralism is arrived at as the most rational economic activity.

Pastoral people of about 10 different social groupings with largely similar dialects predominantly inhabit this region. The population, according to the 1991 census, was about 400,000. In this study, I consider the Karimojong to refer to the groups of the Matheniko, Pian, Bokora and Tepeth.[2]

1.2.2 The Potential for Livestock Production in the Different Agro-Ecological Zones

There are seven agro-ecological zones that have been identified by a recent study (see figure 2, Fintecs Consultants, 1997). These zones are characterised on the basis of their agro-ecological characteristics, including climate, soils and altitude. They determine the main agricultural activities (especially the dominant food and cash crops) as well as other forms of land use such as animal keeping. The potential for livestock production in these different zones is described as follows:

1 The Teso Zone receives precipitation between 1000 mm to 1500mm, with maximum temperatures ranging between 27.5°C and 32.5°C and minimum temperature averaging 17.5°C. The types of soil range from vertisols to Ferralitic and Lithosols. The Ferralitic soils are mainly sandy loams and clay loams. The southern pastoral zone comprises vertisols, young, shallow soils susceptible to erosion. The eutrophic soils on volcanic ash and humic ferralitic soils offer a high to medium productivity. The vegetation types are moist *combretum* savannahs, *Butyrospermum* and grass savannahs. The grasses include *Hyparrhenia* spp., *Panicum maximum, Themeda Triandra, Setaria aequlis* and *sporobdus* spp., and as short grassland are ideal for grazing. It covers the districts of Soroti, Kumi and Katakwi.

2 The Banana Coffee Zone receives precipitation between 1000 mm to 1500 mm, with maximum temperatures ranging between 25°C and 30°C and minimum temperature averaging 15°C. The soils are predominantly ferralitic and are very fertile in areas with pediment soils of the Buganda catena. The vegetation is mainly forest/savannah mosaic. The pastures are suitable for intensive grazing because of the high vegetation biomass production dominated by tall grassland. This type of grassland matures very fast, leading to poor quality pastures as the dry season approaches. The dominant grasses include *pennisetum purpureum, Hyparrhenia spp.*, with the herb layer including *Panicum maximum, Brachiaria spp., Cynondon dactylon, Setaria sphacelata* and *Chloris gayana*. It covers the districts of Iganga, Jinja, Mukono, Kampala, Mpigi, Masaka, Kalangala, Kibale, parts of Kiboga, Kabarole, Luwero, Hoima, Masindi, Rukungiri, Bushenyi and Bugiri.

3 Banana-Cotton Zone receives precipitation between 1000 mm to 1250mm, with maximum temperatures ranging between 27.5°C and 30°C and minimum temperatures ranging between 15°C and 17.5°C. The types of soil are ferralitic and mainly sandy loams. This zone falls under the *Combretum, Terminalia, Butyrospermum* savannah. This area has moderate biomass production and predominant grasses include *Hyparrhenia rufa, Panicum maxi-*

mum and *Setaria sphacelata.* It covers the districts of Tororo, Pallisa, Kamuli, parts of Luwero, Masindi, Mbarara and Ntungamo.

4 The Northern Zone receives precipitation between 1000 mm to 1500mm, with maximum temperatures ranging between 27.5°C and 32.5°C and minimum temperature averaging 17.5°C. The soils comprise mainly Ferruginous tropical soils, lithosaols and hydromorphic soils which are not so productive. Vegetation comprises mainly *Butyrospermum* and moist *Combretum* savannah. The herb layer is mainly comprised of *Hyparrhenia spp., Setaria spp., Panicum maximum, Adropogon gayanas, Brachiaria spp.*, and *Sporobolus spp.* It covers the districts of Lira, Gulu, Apac and Kitgum.

5 West Nile Zone receives precipitation between 1000 mm to 1250 mm, with maximum temperatures averaging 30°C and minimum temperature averaging 17.5°C. The soils are mainly undifferentiated vertisols comprising of the fine clays that swell during wet season and crack during dry season. A big part of this area has sandy loams. This is a sub-humid zone with vegetation associated with the moist *Butyrospermum-Combretum-Terminalia* grassland. The grass layer comprises *Hyparrhenia spp., Panicum maximum, Adropogon gayanas, Brachiaria spp., Setaria sphacelata, Themeda Triandra and Sporobolus pyramidalis.* It covers the districts of Moyo, Arua, Nebbi and Adjumani.

6 The Montane Zone receives precipitation between 1000 mm to 1500 mm, with maximum temperatures averaging 23°C and minimum temperature averaging 10°C. The soils range from non-hydromorphic organic soils in the mountains to eutrophic soils on volcanic ash and humic ferralitic soils. This zone falls in the high altitude moorland and heath, with high altitude forests and forest savannah mosaic. In the areas above 1750 m above sea level, there are no known grasses of value to livestock. However, valuable grasses in moist savannah mosaic occurring below 1500 m above sea level include *Pennisetum clandestinum, Digitaria abyssinica, Cynondon dactylon* and *Panicum maximum.* They cover the districts of Mbale, Kapchorwa, Kisoro, Kabale, Kasese parts Bundibugyo and Kabarole.

7 The Pastoral Zone receives precipitation between 500 mm to 1000 mm, with maximum temperatures ranging between 27°C and 30°C and minimum temperature averaging 15°C. In the North East sub-zone districts of Kotido and Moroto, the vegetation is associated with dry Acacia-Combretum-Terminalia. The herb layer comprises *Hyparrhenia, Setaria, Themeda, Chrysopogon and Sporobolus.* In the south western sub-zone districts of Rakai, Masaka, West Mpigi and most parts of Mbarara the vegetation comprises dry Acacia Combretum grassland. The herb layer comprises Cymbopogon, Nardus, *Themeda Triandra, Brachiaria spp., Panicum maximum, Chloris gayanas* and *Loudetia kangerensis.*

In Uganda livestock production is possible in all the zones, although the pastoral zone followed by the Teso zone have a comparative advantage over the rest of the zones for extensive livestock production. The predominant system of grazing in the traditional sector in these areas is mainly nomadic pastoralism. In the south west pastoral sub-zone, commercial ranching was introduced in the 1960's and is still carried out alongside nomadic pastoralism. The Banana coffee zone and Banana cotton zone have a comparative advantage for intensive livestock management, and in these areas livestock rearing and crop cultivation is practised.

Notes

1 Mamdani, *et al*, 1992, 'Karamoja: Ecology and History,' CBR Working Paper No. 20, Kampala,(p. 2-7) present a detailed description of the ecological zones and variations of rainfall in Karamoja. They give an account of how factors determining soil formation in the different ecological zones in the region contribute to the present ecological differences and variations in Karamoja. See also Dyson-Hudson, 1966, Karimojong Politics (p. 30-32).

2 Other writers like Dyson-Hudson 1966, Lamphear, 1976, and Baker, 1975 have spelt out that the group called the Karimojong comprises the sections of the Matheniko, Pian and Bokora – all of whom live in Moroto district, and that they are distinctly different in customs from say the Jie of Kotido district. However, as will be shown later, the Tepeth are now accepted as Karimojong by the other Karimojong. They now even speak Ngakarimojong (the language) and engage in other Karimojong customs like belonging to the Karimojong clans and practicing initiation.

CHAPTER TWO

Commercial Livestock Ranching in Uganda

2.1 Background

The Ankole is the name the British colonial administration gave to the kingdom of Nkore in 1901. They are located in southern and southwestern Uganda in an area peopled by the Banyankore (singular: Munyankore), whose language is Runyankore. The Ankole Kingdom has a long history dating back to the 15th century, thus surviving the colonial rule, but was abolished, along with the three other Ugandan Kingdoms in 1967 by Prime Minister Dr A. Milton Obote. However in 1993 the Ugandan President Yoweri Museveni restored the kingdom, since the Ankole believed to be his people.

Traditionally, the Ankole lived an agro-pastoral life, with a short migratory pattern around their homesteads. Milk and meat constitute important part of their diet often drank when it is sour and known as called *moursik*. Cattle was and still is the focus of social life, considered a status symbol as well as a depository of wealth also sold in the market to meet household needs or for investment in trade and other business interests.

However, Ankole people have been subject to great social, economic and political transformation. The transformation process took three dimensions: i) herd concentration as the wealthy and powerful pastoralists invested in cattle fattening and selling for higher prices. ii) Some wealthy pastoralists are able to diversify the agro-pastoral activities and became involved in commercial farming and trade. iii) The expansion of large-scale ranches and farms denied the majority of the traditional pastoralist's free access to grazing resources.

This case study focuses particularly on the evolution of pastoral transformation among the Ankole people, with special reference to the expansion of Government-sponsored ranching schemes. It traces this process from the colonial through the per-President Yoweri Museveni Government until today. The case study signifies how these ranching schemes affected the Ankole people as well as the political and power structures that sustain them.

The livestock sub-sector in Uganda is replete with development policy interventions undertaken for reasons, in a manner and with timing, seldom subjected to interrogation. The majority of these interventions are justified mainly for ecological considerations about range-land degradation and the improvement of the efficiency of utilisation of range-land resources. These policies are seldom designed to unravel any underlying social and political ramifications. In the majority of cases they are more useful in explaining the various orientations of these policies. Policy interventions in the livestock sub-sector have been changing over the years, largely as a result of the perceptions of the government in what constitutes the major obstacle for the development of livestock production. This changing policy terrain and its significance is seldom captured in the literature on livestock development in Uganda.

A concrete examination of this changing policy terrain is significant in helping to avoid past mistakes. Post-colonial governments pursued the policy of commercial livestock ranching without ever questioning the contradictions underlying the assumptions made at the time the ranching schemes were established. Similar arguments used to justify the establishment of the ranching schemes during the colonial period continued to be echoed much later to justify the success of the ranching schemes. For example, Kanabi-Nsubuga (1984: 338), writing about the success of the Ankole/Masaka ranching scheme, argued:

> The livestock industry in most African countries is based on a state of traditional pastoralism, in which stock are individually owned, but land is owned and grazed communally. In such circumstances, there is no incentive to any individual to invest in the improvement of the land resources.

Until 1986, official government policy considered the setting up of commercial livestock ranching schemes based on an individualised land tenure system as the most feasible way to transform livestock production. Even when overriding resource use constraints became apparent in attempts to transform livestock production through commercial livestock ranching schemes, immanent resource use constraints were explained away by referring to bottlenecks originating from the traditional livestock production sector (Republic of Uganda, 1988). Others considered that these constraints arose from the new livestock production systems intended to replace the traditional ones. The new systems were seen as creating new forms of interests, which configured with the old livestock production systems to inhibit significantly the regulation of resource use, even after undermining the traditional systems.[1]

It was usually presumed that the traditional pastoralists, in whose midst these large-scale commercial livestock ranching schemes were established would, through the effect of demonstration, benefit from the developments in the livestock production sector (Raikes, 1981). The experience of government-sponsored ranching schemes is all too familiar: the traditional cattle keepers were excluded from such schemes right from the start. Many turned into squatters on the ranches; others remained in the vicinity continuing to encroach on the pastures enclosed by the ranches. This led to multiple resource use problems, including among others, over-grazing. (Republic of Uganda, 1988).

When the NRM assumed power in 1986, the plight of landless cattle keepers who were squatting on both private ranches and on the ranches of government ranching schemes, were considered a bottleneck to livestock production. It was not as significant as the conceptualisation and implementation of this very policy of commercial ranches. While addressing the National Resistance Council on 22 August 1990, President Museveni said:

> Government moved in (by setting up a commission of inquiry) on this issue (of ranching schemes) because it has always been troublesome. People have been fighting ever since, maybe also the question of violence has always been there, it has never stopped ... it has paralysed the whole animal industry in that area. We are really missing a lot by this paralysis because the ranchers bring cattle; the cattle die because the squatters introduce ticks. It is endless, we are loosing a lot of money, the country is loosing and I cannot be a party to this both for the issue of social harmony in the area and also for developmen.[2]

The grounds on which the NRM justified intervention of the government ranching schemes ranged from philanthropy (the lack of social harmony between squatters and ranchers) to economic considerations (the loss to the country as a result of the paralysis). The intervention was also justified on the grounds of equity. President Museveni told a public rally in 1990 that there was no reason for Uganda to have homeless people for the sake of acquiring ranches for a few ranchers who were not using the land optimally.[3] The NRM considered the policy of developing the livestock production sector through the establishment of ranching schemes as flawed from the very start.[4] Its alternative strategy – repossessing the former government ranching schemes and restructuring them to make available land for allocation to former landless cattle keepers – is based on the same assumptions as that of the policy it criticises. Private land ownership in the range-lands is a requisite condition for the transformation of the livestock production sector.

This means that it is very significant to try and understand the specific goals and objectives that the various governments sought to develop the livestock sector. While the role played by the state is important in understanding the rationale for policy interventions and, to a great extent, the implications these policies have on the allocation and use of available scarce resources, the possible responses of those affected by the respective policies also need to be examined. This is because sometimes, even well intentioned policy interventions in the livestock sector, have resulted in undesirable consequences both to the cattle keepers and the range-lands (Raikes, 1981). This suggests that both the overt and sometimes the covert objectives and intentions of the policies for the development of the livestock sector need to be investigated.

The fact that these policy interventions contribute to changes in the resource availability among different stakeholders, implies that they cannot be considered 'neutral'. As far as this study is concerned, the reasons that have been given for certain policy failures are not 'neutral' either, and therefore, also need to be unpacked. Livestock sector development policies, how they are implemented, how those who are affected by these policies respond to them, and how the problems and failures of these policies are explained, have implications that are political in nature. And these implications of policy are influenced by the general political economic tendencies at a specific point in time.

The fore-going prognosis suggests that because the primary objectives for developing the livestock sector have been different, there has been a variance in the definitions of what the livestock production problem constitutes. As a result, there has been a problem of what the best possible implementation strategy for livestock policies would entail. There have also been divergences in explaining the failures of these policies. We agree with Raikes (1981) that sometimes the problem is not simply the ineffective implementation of a basically correct policy. Nor would it simply be one of incorrect policy choices arising from lack of information or incorrect assumptions. Even if these defects could be overcome, major problems would still remain because of conflicts of interests and aims. Wherever there has been a policy intervention in the livestock production sector in Uganda, it has had to contend with vested interests by the stakeholders. They have, to a great extent, shaped the manner in which the respective policies are played out.

This predicament is captured in the statements made by President Museveni during the debate in the NRC on the issue of the ranches. President Museveni said:

> It is not a question of whether we love the ranchers or we love the squatters, the issue is that really the government (then) is the one, which brought this confusion between the two peoples. The old governments are the ones who introduced this conflict because they introduced a policy without looking at all the angles of the problem … now, on one hand the ranchers are right because they were given some ranches, and they were given leases, so they say 'this is our land'. On the other hand, the customary

rights of the squatters were not respected because those people, before the modern tenure system came, were living there.[5]

This study contributes to the discussions on the problems of policy interventions in the development of the livestock sector. Since the colonial period, there have been glaring similarities in the leitmotif of the major livestock development policies in Uganda, although the specific orientation of such policies has largely been influenced by the character and nature of the state. This study engages the ramifications of the various livestock development policies, in order to determine the basis upon which the current plight of the pastoral sector can be better understood.[6]

2.2 The Development of Commercial Livestock Production in Uganda

2.2.1 The Colonial State and Livestock Development

The development of the livestock production sector in Uganda dates back to the beginning of the 20th century when Uganda became a British colony. The concerns of the colonial state centred on increasing beef production for both the domestic and export markets. Uplifting the welfare of traditional cattle keepers, it was presumed, would result from the slipover effect of developing commercial livestock ranching. As early as 1904, exporting cows or heifers without special permission was prohibited by notices in the gazette of June 1st, under the Uganda Custom Consolidated order of 1904, section 10, dated 13th May 1904.[7] Although the cattle population of Uganda was estimated to exceed 2,714,000 by the end of 1951, compared to an estimated 2,394,000 five years earlier in 1946, the demand for meat in rural and urban areas was far higher than its supply. The price for cattle for slaughter was very high. The increase in the cattle population of 320,000 in a five year period was considered satisfactory. But in reality the demand for slaughter stock far exceeded the supply during the whole year. Regrettably, there was a reduction in the flow of cattle passing through organised primary stock markets in the producing areas. This was a partial reversion of the barter system of the 1930's due to the scarcity of consumer goods in rural areas. As a result, in 1951, the supply of slaughter stock to urban centres of Uganda was inadequate throughout the whole year.[8]

By the end of the 1950's, it was quite clear that there was an over-riding need to increase the output of livestock products on the domestic markets in the British protectorate of Uganda. However, the colonial government identified an important obstacle to the expansion of meat production from the pastoral sector. According to the colonial state, increasing the percentage off-take of animals presented a sociological problem in the following sense. Pastoral people considered their cattle as a source of wealth, pride and prestige, and therefore restricted their off-take of cattle to the market. Cattle provided food for the pastoralists, as their diets mainly constituted milk. Female calves were preferred more than males, although the latter would provide a useful source of slaughter animals.

Since the prospect of importing large number of livestock was not good, the only way the colonial administration could increase meat production was by increasing carcass weight, which were very low on African holdings. Increasing the weight of stock bred for the market, would require improving range management and animal husbandry. Such an improvement would take a long time to bring about. For these reasons, it was considered unlikely that large increases in meat production would occur quickly without any fundamental changes in the livestock production system (UNDP/FAO, 1967).

It was because of the above that the colonial government chose to develop rapidly the livestock production sector. To be able to do so as rapidly as possible, the authorities had to make a choice. The decision they had to make to develop the livestock sector was to isolate the traditional pastoralists. By 1957 the authorities of the colonial government had identified the following factors as obstructions to the increase of livestock production.

a Extremely poor husbandry and management levels (I.e. Poor calf management led to malnutrition and unnecessarily severe reactions to East coast fever [ECF]. Associated disease conditions led to losses exceeding 200,000 annually) and

b Importation of cattle and the meat of carcasses into Uganda from the Kenya Meat Commission totalling 4,213 cattle on the hoof and 769 tons of beef.

By independence Uganda's livestock production was still in deficit. The East African Livestock Survey of 1963, funded by the World Bank, predicted that by 1971 Uganda would be a net importer of beef. To UNDP/FAO, commercial beef production became the only viable solution. The Government of Uganda planned extensive expansion of its livestock industry, which received massive support from the World Bank mission Report,[9] as well as from the report of the East African Royal Commission.[10]

It was predicted that although per capita consumption of meat and milk was very low in the 1960's, this low level of consumption per head would not be maintained until 1970. There were also economic reasons for expanding livestock production – it was likely that there would be substantial increases in the prices of meat products, which sold at considerably less in East Africa, than on the World Market. This provided an opportunity to look into the possibility of diversifying export to include beef (UNDP/FAO, 1967: 13). The major exports of East Africa, coffee and tea, were faced with declining terms of trade starting in the 1960's. Sisal and Cotton were threatened by cheap synthetic substitutes. It was noted in the UNDP/FAO report:

> It is concluded that East Africa should discontinue its present dependence on these products ... meat does not face the prospect of competition from synthetics which cloud the future of cotton, sisal and wool, and also, as consumers' incomes increase, their expenditures for meat tend to rise more rapidly than for other foods. The market situation for meat is rather complex, but definitely more favourable (UNDP/FAO, 1967: 16-17).

To the colonial government, and later the independence government, it made all the economic sense to develop commercial livestock production in Uganda rapidly. The major donors at the time, the World Bank, the UNDP and the FAO were very positive about the programme. The independence government, badly in need of foreign capital development, embraced the idea wholeheartedly.

2.2.2 The Individualisation of Rangeland Tenure

Apart from the socio-cultural constraints on the development of livestock production (through the expansion of beef production) identified by the colonial government, there were also other constraints arising from traditional institutions of resource management, especially the land tenure system. It was noted in the UNDP/FAO report:

> Livestock improvement will depend upon the adoption of improved methods of animal husbandry by the majority of stockowners. These methods in turn necessitate enclosure to control stock movement, and prevent the spread of diseases; tick control by dipping and spraying, and the development of improved water supply, in the ideal case in the form of permanent installations and reticulates supplies. It will be obvious that the form of land tenure in force in any area must virtually affect all these practices. The man who wishes to grow a tea bush or a coconut tree on a plot of suitable land half a mile from his house, with the farms of many neighbours in between, may still be able to approach nearly optimum yields. But the man who tries to keep a high yielding cow on an un-enclosed plot of land half a mile from his home will not keep it very long: it will either be

> stolen or will die of disease. For livestock improvement, therefore, the right form of land tenure is even more important than it is for arable or permanent cash crops (UNDP/FAO, 1967: 38).

This marked the beginning of the idea of introducing private land holding in the range land areas, through a system of rationalised forms of land use in the name of government sponsored ranching schemes. The idea of creating government sponsored ranching schemes during the 1960's and early 1970's, was seen as a way of putting to better economic use agriculturally marginal areas. The establishment of private rights in property was considered not only an inevitable consequence of the development of society,[11] but more importantly, it was seen as an ultimate goal pursued by all 'rational' individuals including the cattle keepers.[12]

But the land identified as suitable for livestock development was also heavily tsetse infested. The problem of tsetse fly infestation and its effects on livestock production was highlighted in the Annual Report of the veterinary department (on 31 December 1938), which stated the following:

> In Western Province, the widespread trypanosomiasis militates against any marked increase in stock number, and the heavy percentage of carcasses from these areas partially or wholly condemned owing to tuberculosis or cysticercus bovis makes it increasingly difficult to obtain a good market for slaughter bullocks or barren cows. At present the Bahiima, who are nomadic in habit, are the chief stock owners; their custom of moving stocks from place to place, the intermingling of herds at watering places and grazing grounds, and the use of stock in marriage payments do much to facilitate the spread of such diseases as trypanosomiasis, anthrax and black-quarter.[13]

The introduction of private property systems was to be preceded by the containment of tsetse flies. Measures to contain the advance of tsetse flies commenced in 1958 under the Uganda Tsetse Control department, and consisted of forcefully evacuating all the remaining cattle, and slaughtering and hunting all game species in the area in order to starve tsetse flies. This halted the advance of the fly. The continued shooting of game, however, provoked strong opposition from wildlife conservationists. Because the strategy also turned out to be expensive, spraying was adopted, in addition to the creation of a 4 to 5 mile wide and 32 mile long buffer zone separating the area cleared of tsetse flies from the invested areas (Sacker and Trail, 1968).

2.3 Livestock Development in Post-colonial Uganda (1962-1971)

2.3.1 Government's Policy on Livestock Ranching

Uganda achieved its independence from the British on the 9th October 1962. The Obote I government that assumed power immediately after independence, continued with the development of the livestock sector along lines that had been started during the pre-independence period. The Obote I government policy on the development of the livestock production sector was summarised in a speech by the Vice President, who was also Minister of Animal Industry, Game and Fisheries, the Hon. J.K. Babiiha, M.P. At the opening of the RAC on 15 January, 1970, he said the following:

> It is now government's intention to bring a much wider section of the community into economic ranching, and the primary task of the RAC will be to advise we on how best to apply our resources of capital and labour to the land. Not only does the rancher need a fair return for his work and enterprise, and the public expects rapidly a good product at a fair price, but that in establishing new enterprises we must take care to protect existing interests. Good land is our greatest natural asset and while we clearly need to extract from the land the maximum economical returns, we must make quite sure that we pass on this great asset to succeeding generations in at least as good heart as we found it.[14]

The Obote I government policy emphasised the development of commercial livestock production through the establishment of large-scale commercial ranches. It also emphasised the rationalisation of ranching to 'bring (in) a much wider section of the community'. And this is where much of the dilemma was. While the colonial government was clear as to the position of traditional cattle keepers, the Obote I government policies had an ideological predisposition rooted in the common man's charter. The common man's charter emphasised, among others, that economic development, and in this case the development of commercial ranching schemes, should directly benefit the common man in the community. The community that was talked about during the Obote I government later turned out to be a 'narrow' community of mainly the political elite.

Obote I's government policy for developing the livestock sector was also concerned with sustainable utilisation of marginal range land resources. The policy also advocated the need to make use of land economically in the present period, so that it could be available for future use. Sustainable utilisation or even the protection of existing interests never referred to improving traditional pastoralism.

2.3.2 The Creation of Government Ranching Schemes

The Pilot Land Use Investigation Unit

While the idea of the ranching schemes was conceived before independence, it began to be put in operation during the Obote I government. Prior to the establishment of the ranching schemes, the colonial government set out to eradicate tsetse flies that afflicted livestock production in much of the areas that had been earmarked for large-scale commercial ranching schemes. Concurrent to the attack on tsetse flies, efforts were made to develop rational land use for the area when land would become available, and in 1957, a pilot Land Use Investigation Unit (LUIU) of 30 square miles was established in the area. The LUIU was intended to make preliminary studies of pasture utilisation, grazing and fire control, and to investigate techniques for construction of small dams and catchment tanks. Following the establishment of the LUIU, the government approached USAID in 1962 to make a pre-investment survey on the basis of which more detailed plans for ranching development would be undertaken. This pre-investment survey was undertaken between January and April 1963 by a team headed by Dr. K. Gregory (Sacker and Trail, 1968).

The Ankole/Masaka ranching development project (number 617-26-130-033, USAID Loan No. 617-H-004) was based largely on the recommendations of the 1963 USAID livestock survey team, otherwise referred to as the Gregory Report (USAID PLO/T 617-A-14-AA-2-30001, May 1963).[15]

The Establishment of the First Ranching Scheme

The first phase of the project lasting from 1963 to 1973 consisted of the establishment of 62 ranch units covering 1200 hectares. This was carried out in Ankole. Each of these ranching units was scheduled to carry a total of 1000 head of stock by the time of stabilisation in 1974 (Kanabi-Nsubuga, 1984). By mid July 1966, 28 ranches in Ankole had already been allocated to successful applicants.[16] Of these, 12 ranches were ready for occupation by August 1966.[17] The members of the Ranch Allocation Committee (RAC) generally felt that ranch size should be based on an economic unit of cattle, and acreage per ranch would therefore vary between areas according to environment.[18]

At the time when the central government was busy establishing the Ankole ranching scheme in the 1960's, the Buganda government also undertook the establishment of the Buruli Ranching scheme as a pioneer scheme in ranch development for the Kingdom. The ranching scheme in Buruli was developed along similar lines to the one started by the central government in Ankole. 27 ranches were created, out of which the Buganda Land Board leased 26 ranches to individuals, companies and co-operatives. Leasing of ranches started in 1961 and by 1968 all ranches had completed the initial lease period of 5 years.[19]

The Allocation of the Ranches

The Buganda government had begun to allocate ranches for the Buruli ranching scheme and was being undertaken by the Buganda establishment at Mengo. The allocation of ranches in the Ankole Ranching Scheme was undertaken by the central government through the then Ministry of Animal Industry, Game and Fisheries, that supervised the activities of the Ranches Selection Board (RSB). It was hoped that applicants for ranches in the ranching schemes would be carefully selected in order to concentrate the most promising ranchers in the most promising area, in order to obtain maximum advantage from limited capital and limited availability technical staff.[20] We were able to establish that it was only the very first ranchers who were intensively screened before being allocated ranches. Subsequent re-allocation of ranches failed to meet with the strict criteria that had initially been proposed. This saw the entry into the ranching enterprise of a local political elite, which was closely associated with the government in power at that time. Below is an examination of the process of allocation.

The Ranching Policy Advisory Board (RPAB) put in place some guidelines for the Ranches Selection Board (RSB) to follow in the allocation of ranches in the first phase of Ranch Development in the Ankole Ranching Scheme, and included the following:[21]

i The first priority in selection was given to local applicants, defined as all residents of Ankole, plus all Banyankore.

ii Allocations were to be made for only the ranches in the first phase, numbering 28. Unsuccessful applicants were to be reconsidered when more ranches became available.

iii The occupants of the LUIU at the time of the allocation of the first phase ranches were also considered during the selections for ranch allocation. There were six Bahiima families in the areas where the LUIU was established. They were only required to agree to the terms of the leases. They were given priority to retain their present sites if they wished to do so, as long as they complied with the terms and conditions of the lease. As it turned out, they all failed to meet these terms and were displaced.

Applications were invited from the members of the public (in Ankole, this also included Kigezi) who met the above criteria. The RSB had constituted a sub-committee that was to carry out preliminary interviews with ranch applicants before they were finally called for an interview by the complete RSB. This sub-committee collected all factual data about a person's application and would then report to the RSB.[22] During the preliminary interviews, the sub-committee of the RSB did not interview all the applicants. It was not considered necessary due to their credentials. These applicants were mainly prominent politicians from Ankole, who would be among those from whom the actual selection of the ranchers was to be done by the full RSB.[23] Only the ones whose credentials in terms of animal production the RSB was not sure about were to be interviewed by the sub-committee.

It was assumed that the more politically prominent a person was, the better placed that person would be in terms of livestock husbandry. Political prominence was related directly to how closely one was associated with the ruling party, the Uganda People's Congress (UPC). In fact some of these prominent politicians who applied for ranches did not appear before the sub-committee. One such applicant sent an application by telegram indicating (verbatim) that:

> Unable to attend (interviews). I have increased my cattle to 100. Willing to furnish details. Definitely want ranch.[24]

The Permanent Secretary of the parent Ministry of Animal Industry, Game and Fisheries contributed significantly to perverting the allocation process. He wrote to the Secretary of the RSB, informing him that he himself had been informed that Members of Parliament from Ankole were to appear before the RSB sub-committee for interviews. After the RSB checked their applications, they said that he (the RSB Secretary) should use his discretion to exempt some of them from interviews. He argued:

> There may be others, who likewise, after checking their applications, you may consider may be exempt from being interviewed by the sub-committee, in which case these should also be informed accordingly in writing.[25]

It is also important to note that during the first and second ranch allocation by the RSB, there were no clearly stipulated criteria for allocation of the ranches. The first applicants were interviewed by an RSB sub-committee. Then a short-list of applicants would be made whose names would be forwarded to the full RSB. The basis of the two sets of interviews were as follows: to obtain factual information concerning cash and cattle resources; to determine if they really knew the requirements for the commercial ranches; and to attempt to decide if they had the ability and drive necessary to run the ranches.[26] It was at the 3rd meeting of the RSB that guidelines were formulated on the criterion to be followed during the allocation process.[27]

By 1967, there were already complaints that potential applicants had not been given adequate opportunity to apply for the ranches. At its 4th meeting, the RSB noted that all promising candidates of the applications for the ranches in Ankole had already been interviewed. The members of the RSB requested their secretary to ensure that nation-wide publicity be given well before the next allocations of ranches was considered by the Board. The best possible applicant would then be given every opportunity to apply.[28]

Since there were usually fewer ranches than the applicants, it is interesting to note the identities of some of those who were not allocated ranches. Even if they had been given first priority during the proceeding selections, they would have been interviewed again to find out if they were still interested. During the second selection, only 20 applicants were selected for interviews by the RSB on 27 August 1966. 55 applicants were not selected for interviews. This included the only female applicant, Mrs. Joyce Rwetsiba, who was the wife of the then Permanent Secretary, of the parent Ministry. There was also an army Major (in 1966!) who was not allocated a ranch the first time he applied.

There were usually enormous pressures from the members of parliament from Ankole and Kigezi on the RSB to be allocated ranches. A Member of Parliament from Kigezi, and a Deputy Minister of Animal Industry, Game and Fisheries, in his application for a ranch, indicated that he had 250 heads of Ankole cattle and 40 exotic cattle. The Deputy Minister argued:

> Land shortage in Kigezi cannot allow me to keep these cattle, moreover, as it is the exotic cattle which bring me some income, I would like to allow them the available land and move away the 250 Ankole cattle.[29]

As regards the application of the Hon. Deputy Minister above, the Permanent Secretary, Ministry of Animal Industry, Game and Fisheries wrote to the Senior Livestock Improvement Officer as follows:

> I confirm our discussion that this (the Deputy Minister's application) will be included in the list for consideration by the selection Board.[30]

Because of the enormous pressure from prominent politicians, especially Cabinet Ministers and their Deputies, the USAID, which was financing the project, insisted on a loan agreement for the project. Politicians would not qualify for USAID loans for ranch infrastructure development, if they were even selected for ranch allocation. The Permanent Secretary noted:

> The point should be made that the allocation of a ranch to (the deputy Minister) will not qualify him for a loan assistance in view of H.E. the President's ruling contained in memorandum dated 26 April, 1965 addressed to the then Director of Planning. However, (the deputy Minister) as a progressive farmer cannot be denied opportunity to improve his farm for the principle here is that professional politicians do not live on their politics alone, but are permitted to have other interests besides their politics. The restrictions imposed on the acquisition of ranches by Ministers and Deputy Ministers was the result of conditions precedent in the USAID Loan Agreement and to meet this requirement, the RSB should advise the Deputy Minister that should he be allocated a ranch, he would have to develop it from his own resources without loan assistance from USAID.[31]

These conditions set by the USAID set the stage for a major diplomatic row between the government of Uganda and the American Embassy, which we will examine below.

The Controversy over the Allocation of Ranches to Politicians

During the allocation of ranches, there was a lot of pressure on the government from the 1st Parliament of Independent Uganda for Members of Parliament to be allocated ranches. The USAID, on the other hand, was very particular about involving politicians in the business of commercial ranching. It was argued by the USAID that while the members of parliament would be better off developing the ranches since they already had some resources, there would a problem of monitoring the development of the ranches because the state would be heavily involved.

According to Doornbos (1971), the Gregory report, upon whose recommendations the establishment of the Ankole ranching scheme had been based, was opposed to the allocation of ranches to absentee owners because this would prevent it from becoming a model of social change. The Gregory team preferred the establishment of cattle ranching to be a gradual matter. There was interest especially from officials in the Ankole kingdom government as well as high-ranking officials from the Ministry of Animal Industry, Game and Fisheries, for a more rapid progress. They had perceived the scheme as an opportunity for immediate economic gain. The Uganda government had given very little consideration to the impact of the scheme on the standards of living of the people of the area. USAID became reluctant to proceed with the project until issues of social benefit were satisfactorily clarified. Doornbos observed

> The tone of USAID's position indicated a growing suspicion that the issue of rancher selection was more than an abstract consideration and that

> there was an active lobby in Ankole eager to capitalise on any flexibility in the criteria for eligibility for ranch ownership. Washington's tendency to take a critical position, in turn aroused the sensitivity of officials in the Ministry of Animal Industry, who became impatient at what they viewed as 'America's dilatory tactics'. Official contacts between the two countries were accompanied by mounting tension as the problem was treated in a more and more unequivocal manner ... the growing estrangement of the two governments over this matter generated strong expressions of indignation among Ugandan officials. They claimed that the United States' desire to have a voice in the policy of ranch allocation was unwarranted interference in the exclusive right of Uganda government to implement its development programme as it saw fit.[32]

High-level government meetings were convened to discuss whether the USAID loan for the development of the Ankole/Masaka ranching scheme should be rejected, or if a fresh application be re-submitted. Some radical politicians used this issue to expound conveniently anti-American views. On 2 December 1964, the Director of the USAID wrote an outspoken letter to the Permanent Secretary of the Ministry of Animal Industry, Game and Fisheries, making it clear that the USAID viewed the principle of absentee ownership as a thinly veiled subterfuge in which a pressure group of influential political could acquire ranches.[33] In December 1964, the government's application for the USAID loan was withdrawn, citing (among others matters), that the offensive nature of the USAID letter was tantamount to a breach of diplomatic etiquette. Subsequent discussions and exchanges of communication resulted in an apology by the American Ambassador, who specifically mentioned that the USAID Director did not speak on behalf of the United States Government. The Ugandan Government accepted the apology. Discussions opened on the loan application by the government. The minutes of that meeting went as follows:[34]

Hon. Babiiha: Delaying tactics, not attaching any importance to the loan applications, ranches could be developed with our own resources, without any further delay.

Hon. Odaka: Government of Uganda had already rejected several offers (e.g. AID for Police Housing, Peace Corps, etc.);

Hon. Odaka proposed that the loan application be re-submitted.

Mr. Kakonge: Went to say that he was not particularly happy with the way the 1st and 2nd phase ranches in Ankole and Masaka ranching schemes were allocated;

Mr. Kakonge felt that the 'common man' especially co-operative groups should have received top priority;

Mr. Kakonge suggested that any future allocation of ranches should be handled by a planning commission sub-committee.
Dr. Obote: Also after reading through the list of the people or groups of people, to whom the 28 ranches had been allocated, expressed similar views as those expressed by Mr. Kakonge;
Dr. Obote asked that the list of recipients be forwarded to him.

When the list of the 28 beneficiaries was made public, it became apparent that after all, the USAID's fears had become true. This was a rather embarrassing situation for the government which, following the apology the American Ambassador, decided that the loan application be re-submitted and further ranch selection be carried out in consultations with the USAID. But it was rather too late as many of the ranches had already been allocated to leading Ankole politicians or to people with strong connections with the political establishment.

Ranching Co-operatives

The Obote I government believed the best way the common man would benefit from the establishment of government ranching schemes would be through the formation of co-operatives, joint ventures or partnerships. A deliberate policy was therefore advocated for by the RPAB to encourage the emergence of Ranching Co-operatives and Ranching Public Limited Liability Companies in the Ankole/Masaka Ranching Scheme. Not less than two ranches were to be allocated to registered co-operative societies, while at least one locally formed company was to be considered unless their application were otherwise unsatisfactory. The limited companies were required to furnish a list of names of their directors and members.[35]

In the very first selections, 20 applications were considered by the RSB in April 1965. Out of these, 15 were individual applicants, 9 were applications from companies, and 4 applications were from co-operative societies.[36] Co-operatives and companies were required to submit a formal application for a ranch, a list of their members, names of the elected officials and the details of available cattle and cash. Co-operatives were seen as the only way of getting politicians into ranches without raising many questions from the donors. Two groups of co-operatives, each group containing four societies, withdrew their original eight separate applications, and re-submitted two applications, each application referring to a joint venture of four societies.[37]

The first Co-operative joint venture comprised the following societies: Abahamire Growers Co-operative Society, Mutimagumwe Co-operative Society, Engabo Nyabikungu Co-operative Society and Enganguzi Co-operative society. All registered Co-operative societies submitted a joint application and were advised to choose a new name. The President of Abahamire would be responsible for the joint venture in running the ranch. Each society would have a 25 percent share in the ranch. The President of Abahamire would be the ranch manager, and would live on the ranch because he had cattle of his own and therefore some experience with cattle. The group collected 200 heads of cattle, and contributed Shs. 30,000.00 in cash.

The Reconstitution of the Ranches Selection Board

Apart from the political pressure put on the committee that allocated ranches (the Ranches Selection Board), the composition of this committee was initially not a big issue, especially during the allocation of ranches in Ankole. The RSB then had two representatives from Ankole and comprised the following:[38]

i Its Chairman was a magistrate from Teso Local Administration;
ii The deputy Chairman was a representative from the Public Service Commission;
iii The Minister of Finance in the Ankole Kingdom government was a representative of the Omugabe of Ankole.
iv The County Chief of Rwampara
v The Commissioner of Veterinary Services and Animal Industry,
vi The Chief Veterinary Officer, Buganda
vii The Senior Livestock Improvement Officer in charge of the Ankole/ Masaka Ranching Scheme was the Secretary of the RSB.

After the ranches in the Ankole ranching scheme had all been allocated, the RSB moved to Masaka (Kabula ranches). This is when the composition of this committee became an issue. There was pressure to replace the two Ankole representatives with representatives from Masaka. Changes were made in the composition of the RSB, replacing Mr. S. Kayanja, Chief Veterinary officer from Buganda with Mr. S. Sebowa.[39] The new RSB held its first meeting, its 5th for the RSB, on 20 October 1967. At this meeting Applications for a few ranches in Ankole that had fallen vacant were considered at this meeting, and for the first time, ranches in Masaka.

It was usual that some ranches fell vacant in Ankole. Others were repossessed after their owners failed to meet development conditions imposed on them. Simply replacing some members from Ankole with some from Masaka after the allocation of Ankole ranches had ended, created a need to re-constitute a new Board whenever there were ranches from Ankole to be considered. This meant that the membership of the Ranches Selection Board would remain indeterminate, since there would always have to be separate RSB meeting for Ankole and Masaka. This was an awkward situation. The Secretary of the RSB observed:

> It might be taken to imply that ranches in Ankole are allocated to Banyankore and ranches in Masaka to Baganda. This is not necessarily intended, as applications are specified as open to individual Ugandans, co-operatives or Companies. Large committees tend to be unwieldy, and accordingly, I would propose to you that the RSB composition be altered by replacing one of the Ankole representatives by one Masaka representative. Thereafter, this board would attend to any business in hand.[40]

By 1 July 1969, a new allocation committee, which would be called the Ranching Advisory Committee (RAC), was constituted to replace the RSB.[41] The members of the new RAC were announced on 10 July 1969 to replace the RSB.[42] The Ranches Advisory Committee had representatives from all regions and government departments. The RAC had broader terms of reference. The functions of the RAC were as follows:[43]

a The RAC was concerned with the examination, approval and rejection of applications for ranches, instead of the Ankole/Masaka RSB.

b The RAC was concerned with all ranching schemes throughout the country. It was considered to be a more 'national committee' than the RSB which was confined to the Ankole ranches. Even in name it was called the Ankole/Masaka Ranches Selection Board and its terms of reference confined its operation to Ankole and Masaka.

2.3.3 The Terms and Conditions of Ranch Development

Evaluation of Perfomance of Ranchers

The RSB wanted to evaluate the performance of the ranching schemes by 1974. It was however, the view of the Ministry of Animal Industry, Game and Fisheries that the RSB was not suitable for evaluating the ranching schemes because evaluation was both a professional and technical job. They argued that the RSB had professional but private people on the board, who were selected for their abilities in other fields; the RSB was not competent to carry out the evaluation because it was outside its terms of reference. Arrangements for evaluation would be made between the commissioner of veterinary services and animal industry and the permanent secretary of the Ministry of Animal Industry, Game and Fisheries.[44] Despite this, RSB continued a day-to-day assessment of the performance of the ranching schemes in terms of meeting the development conditions imposed on the ranchers.

The ranches were offered on lease to farmers who were selected for capacity, business acumen, educational background, integrity, experience, or a satisfactory combination of those requirements. Individuals, co-operatives or similar bodies that would employ a ranch manager of a similar calibre would also be considered for allocation of a ranch.[45] After the RSB had approved of an applicant's ability from a series of interviews, the RSB would then recommend ranch allocation on a conditional lease. Those allocated a ranch was given a period of three months from the time of allocation to entry, a period within which they were required to sign a lease and pay the conveyance fees and ground rents. The RSB would review the leases after a few years, after which it would grant a final lease approval. If the RSB was convinced that the applicant had furnished the RSB with sufficient information on the basis of which a decision could be made, the applicants would be asked to come to the interviews.

After a ranch was allocated, the RSB required that any applicant be in active occupation not later than a specific date. The number of animals not be exceeded on the ranch was specified. In case the number of animals exceeded the specified amount, special written authority was required from the officer-in-charge of the ranching scheme. Applicants had to indicate their acceptance immediately, and had to pay rent, stamp duty, and conveyance fees to the district surveyor before occupying the ranches.[46]

Ranchers Who Voluntarily Gave up Ranches

Although by September 1968, there were no ranches available for allocation in Ankole,[47] ranches regularly fell vacant. Some of those who had been allocated ranches would turn down offers, others would change their minds, or even fail to fulfil the terms and conditions of entry. (These were things they were required to do before they could effectively occupy the ranches.) Some could not meet the development conditions once they were in effective occupation of the ranches. The level of professionalism that was expected from those applicants benefitting from the ranching schemes could be seen by the extent to which some individuals and organisations withdrew, even after they had been allocated the ranches. One case in point is the Ankole Kingdom government. With regard to an allocation which they had received, a representative of the kingdom government informed the Secretary of RSB as follows:

> I regret to inform you that after reconsidering the possibilities and impossibilities of a government owning ranches and expenses involved therein at this time when there are a lot of things to be done with the funds available, we decided to give up these two ranches which could be taken over by the people.[48]

At that time it was acceptable that individuals or organisation could turn down offers of ranches (when they applied for them in the first place) after discovering that they were not able to manage them properly. This was not to be case in the late 1970's and 1980's.

Eviction of Ranchers for Non-fulfilment of Development Conditions

At its 9th meeting, the RSB considered lease offers for some ranch owners, especially when their performance was to be evaluated. A Mr. E. Khunyirano, by August 1968, had been occupying ranch number 4 in Ankole for five years, and had taken no steps to build a dip until after the Uganda Land Commission was advised to give him notice to quit. The work on the dip had commenced before the notice to quit was served, but was completed after the notice was served. The RSB noted that he should receive compensation for the dip if the notice to quit remained in force. Khunyirano, in a letter of 31 August, 1968, addressed to the Commissioner of Land and Surveys and copied to the Permanent Secretary in the Ministry of Animal Industry, Game and Fisheries and the Commissioner of Veterinary Services, appealed for the lifting of his notice to quit the ranch. However, the RSB noted the following on the appeal:

> After careful consideration, the Board was of the opinion that notices to quit were not intended as a method of replacing the rancher concerned. If this notice to quit were lifted, other poor ranchers would take no action to improve until a similar notice was served. Accordingly, the Secretary (RSB) was requested to advise the PS, MAIGF that in the opinion of the RSB, the notice to quit served on Mr. Khunyirano, ranch number 4, Ankole should remain in force and that in due course Mr. Khunyirano should leave the ranch. It was further noted that it would be desirable in future if any rancher who was being considered for notice to quit should be restrained from building any improvements until a final decision on whether or not to serve a notice had been taken and the notice actually served if so decided.[49]

Many other individuals who had been allocated ranches lost them because they failed to fulfil terms and conditions of entry imposed on the ranches. Take the following example:

> I refer to your visit to my office on 14 March 1968 when I informed you I required an early decision as to whether or not you wished to withdraw from the ranching scheme, and when I informed you that your ranch was ready for occupation. Also to my letter of 19 January 1968, setting out the terms and conditions for entry, which you have failed to fulfil. I am to inform you that the offer of the ranch number 11 made to you by the RSB in my letter of 19 January 1968 has now been withdrawn, and no further offer of a ranch will be made to you.[50]

In Masaka, inefficient ranchers lost their ranches due to failure to put up development activities on the ranches. Take the case of ranch number 17 in the Masaka ranching which was allocated to J.K. Sendagire and Partners on 28 October, 1968, after agreeing to the usual conditions including formalisation and registration of a company as well as to occupy the ranch within three months. The Secretary of the RSB, Marples, noted that:

> They (Sendagire and Partners) wrote on 24 February 1969 asking for this period (of three months) to be extended for a further three months. This extension was granted. However, no development whatsoever has occurred to date. I wrote to them on 24 April 1969 inquiring if any of the conditions necessary prior to the entry had been fulfilled. The letter was returned by the Post Office marked, 'addressee unknown'. I personally checked the address on both letter and envelope and confirm that they were correct. They have made no attempt to sign a lease or pay rent, stamp duty etc. I very strongly suggest, therefore, that the allocation be withdrawn immediately.[51]

A similar case involved ranch number 14, Masaka ranching scheme, which was allocated on 28 October 1968 to Uganda Food Supply. The Secretary noted the following:

> I wrote to them on 30 January 1969 asking if the conditions prior to entry had been fulfilled, and again on 24 April 1969, seeking information. I have had no word in reply. They have sought no extension of the period of three months allowed from the time of allocation to entry. Again they have made no attempt to sign a lease or pay any fees. They well no longer exist as a functioning organisation. Again I strongly recommend that the allocation of the ranch be immediately withdrawn.[52]

The Secretary RSB had indicated to the Permanent Secretary of the Ministry of Animal Industry, Game and Fisheries that he would be glad if the Permanent Secretary would notify the two ranch owners of the Masaka ranches numbers 14 and 17, that their allocations had been withdrawn. The Permanent Secretary could also authorise the Secretary RSB to do so.[53] No action could be taken because by 1 July, 1969, a new allocation committee, which was to be called the Ranching Advisory Committee (RAC), was being constituted to replace the RSB.[54]

Many ranch owners in the ranching schemes that were developed during the second phase had problems putting in place the basic ranching infrastructure including perimeter fencing. This was the situation in the Singo ranching scheme where the commissioner of veterinary services warned ranchers in

Singo who had not erected perimeter fencing for their ranches. Ranch number 17 in Singo ranching scheme was warned as follows:

> On several occasions, I have appealed to you that the perimeter fencing of the above Farm had gone to pieces and being situated on the outer boundary of the scheme, not only the wild animals that roam in and about, but also the nearby 'Bahiima' graze their herds there, thus bringing in uncontrollable diseases, resulting in heavy casualty.[55]

The RAC at its meeting of September 1971 considered further the problem of identification and eviction of unsatisfactory ranchers in the government ranching scheme. Details of unsatisfactory ranchers were to be submitted to the Permanent Secretary or Commissioner, Veterinary services and Ministry of Animal Industry, Game and Fisheries by the Regional Veterinary Officers, Western and Buganda. The report would then come to the RAC, which would then advise the Minister of Animal Industry, Game and Fisheries on the course of action to be taken. When subsequent notices to quit were made by the Minister of Mineral and Water resources, it will be within a period of three months from the date of serving the notice, and will not be subject to any further review.[56]

By the end of 1971, ranchers were required to submit returns of stock to the RAC on a regular basis, according to which their performance would be evaluated. This was the case for owners of ranch number 6 Masaka, to whom the Secretary of the RAC wrote:

> Ranch number 6; Masaka was allocated to you in 1968 under the terms of the leasehold agreement. You then signed, you agreed to achieve a minimum stocking rate on the said land of 200 heads of cattle, of which not less than 100 shall be females of breeding age within the initial terms of two years. According to the most recent returns of stock on the scheme I have received, you have a total of 59 herds of cattle on the ranch, which number include 10 breeding females only. I must therefore, call upon you to show cause why your leasehold agreement with the Uganda Land Commission should not be terminated.[57]

Although the Rwampara Ranching Co-operatives had occupied two ranches in the Ankole scheme, number 18 and 19, even after initial protests during 1967 and 1968 by the then RSB,[58] in 1972 they were unable to hold onto both ranches any longer. The RAC recommended the termination of their tenancy on one ranch, number 18, because:

> *first*, the total number of cattle carried and particularly the number and quality of the breeding females carried, had not been sufficient to justify

> the occupation of the two ranches;
> *secondly*, the level of management achieved had not been capable of efficient utilisation of two ranches and in particular the level of tick control achieved over the area of the two ranches was inadequate and presented a threat to adjacent ranchers with susceptible stock; third, facility development on the two ranches had been minimal.[59]

2.3.4 Resource Degradation on the Ranches

Apart from failure to fulfil terms and conditions imposed on the ranches allocated by the RSB, there were also natural resource management problems on the ranches. These problems began to appear on some ranching schemes, as early as the end of 1971. In the Buruli ranching scheme, for example, the conditions on ranch number 6, as observed by the Commissioner of Veterinary Services while touring the scheme, were described as follows:

> The pasture was badly over-grazed. Considering the season of the year when there should be plenty of grass for grazing, the cattle were observed to be biting earth instead of grass, since several patches of land were almost bare. This was caused by over-grazing. The ranch manager had revealed that the ranch had two square miles of grazing land available, with a total number of approximately 600 herds of cattle, which was more than double the number of animals the area is expected to carry.[60]

Over-stocking and over-grazing caused much of the resource degradation on the ranches. This situation continued throughout the 1970's and 1980's. As a result of the over-stocking which led to over-grazing on many of the ranches in the ranching schemes, land degradation and ecological changes began to take place. Over-stocking on some ranches in the ranching schemes has caused soil erosion creating patches devoid of vegetation, particularly around watering places and night paddocks. This led not only to loss of the already low nutrient levels but also increased silting of valley tanks and valley dams.

By 1988 the nutritionally better quality pasture grass species comprised only about 30 percent of the vegetation cover. These better quality grass species included *Brachiaria spp., Panicum spp., Chloris spp.*, and *Cynodon spp.* They had been under threat of being ousted out of much of the ranching scheme by Acacia tree *spp.* and the aggressively competitive low quality grass. There was an increase in highly competitive species such as Acacia and *Cymbopogon*, especially in communal grazing areas. The ranching schemes became invaded in varying degrees by trees, shrubs, thicket and anthills, which provided conducive breeding habitats for tsetse flies. The result was that by the end of the 1980's, there had been a resurgence of tsetse flies in many of the ranching schemes (Republic of Uganda, 1988).

One of the most significant factors that affected water availability on the ranching schemes that much of the cattle corridor lies in the rain shadow of the prevailing winds, and receives between 500 to 1500 mm of annual rainfall with a very high degree of fluctuation in years and locations. The area is also often subjected to severe effects of drought. Rainfall patterns and intensity have changed over the years and has affected the overall water supply and availability. While rainfall patterns and intensity have greatly affected water availability, the amount of water in valley tanks and dams during the 1980's was also highly influenced by water usage on the ranching schemes. The majority of the watering points lacked a proper reticulation of water supply. The cattle keepers often directly watered their animals at the watering points. Some water troughs were constructed at the edges or sometimes inside the water reservoirs. As the ranches fell to disuse, so did the watering equipment. The pumps, troughs, pipes, etc. had been badly neglected.

Apart from the operational capacity of some of the valley tanks, which was considerably reduced due to the poor design and silting, many watering points were neglected, vandalised or affected by the lack of fences if they existed at all. Fences where important because they barred animals from moving beyond a certain point into the water. This created over-stocking around available watering points resulting in high levels of silt gathering in valley tanks and dams. This was caused by heavy soil erosion around these watering points. The situation was worsened by a lack of de-silting equipment and the uncontrolled watering of large herds of cattle that belonged to squatters.

2.4 Resistance to the Establishment of the Ranching Schemes

Apart from the stringent development conditions, there was another dimension to the problems of ranch development that the ranch owners had to contend with – that of the traditional cattle keepers who had been alienated from their grazing lands to establish the ranching schemes. The focus by the governments since the colonial period on the development of commercial livestock ranching was in total disregard of the traditional livestock production sector. It is important to note that even before the ranching schemes were started, contestation existed between prospective landowners and nomadic cattle keepers. In Ankole, for example, the enclosure of land for cattle keeping or agricultural purposes was already an issue of great concern in 1963. The

Ankole Kingdom Land Board was prompted to instruct all county chiefs to stop land enclosures until the Kingdom Land Board approved it by issue of a certificate of occupation.[61]

In Northern Uganda where government ranching schemes had been earmarked for development during phase two, a lot of resistance to ranching schemes was registered. The Ministry of Veterinary Services and Animal Industry tried to develop a ranching scheme in Akokoro areas, but the local people refused to quit the area. The Commissioner of Veterinary Services and Animal Industry noted that:

> There is some land which would apparently not be occupied and which this department would have liked to develop as original ranching scheme, but there seems to be some objection to it.[62]

The Department of Veterinary Services and Animal Industry did not force the local people to quit the area. The officer in charge, Maruzi noted:

> I have been advised to develop, for beef ranching purposes at Akokoro, that parcel of land which arouses no bitter feelings among the people of the area.[63]

The foundation stone for the Maruzi ranching scheme was laid in the northern side of the peninsula in early 1968. However, due to political pressures,a site of 2000 hectare was established in the southern part of the peninsula. This is the area where President Obote came from, but it is also the home of many communal grazers. The establishment of the ranches in Maruzi would greatly have reduced dry season grazing areas along prime grazing lands around the Lake Kioga shores.

The support that the central government would have provided for the functioning of the ranching schemes in Buganda (namely: Masaka, Singo and Buruli ranching schemes) was greatly affected when tensions developed between the central government and the Buganda Kingdom government. The abolition of kingdoms in 1966 led to tensions between the central government and the Buganda Kingdom government. One Deputy Minister in the central government who hailed from Kigezi had been allocated a ranch in Masaka. He noted the following:

> I would like to be considered for one of the Ankole ranches rather than the Masaka ones. This is purely on political grounds. I am sure you are aware of the Baganda hostile attitude towards government members as a result of the Constitutional changes. I would not like to take a risk in this matter.[64]

The problem of landless squatter pastoralists on the ranches will be discussed in greater detail later.

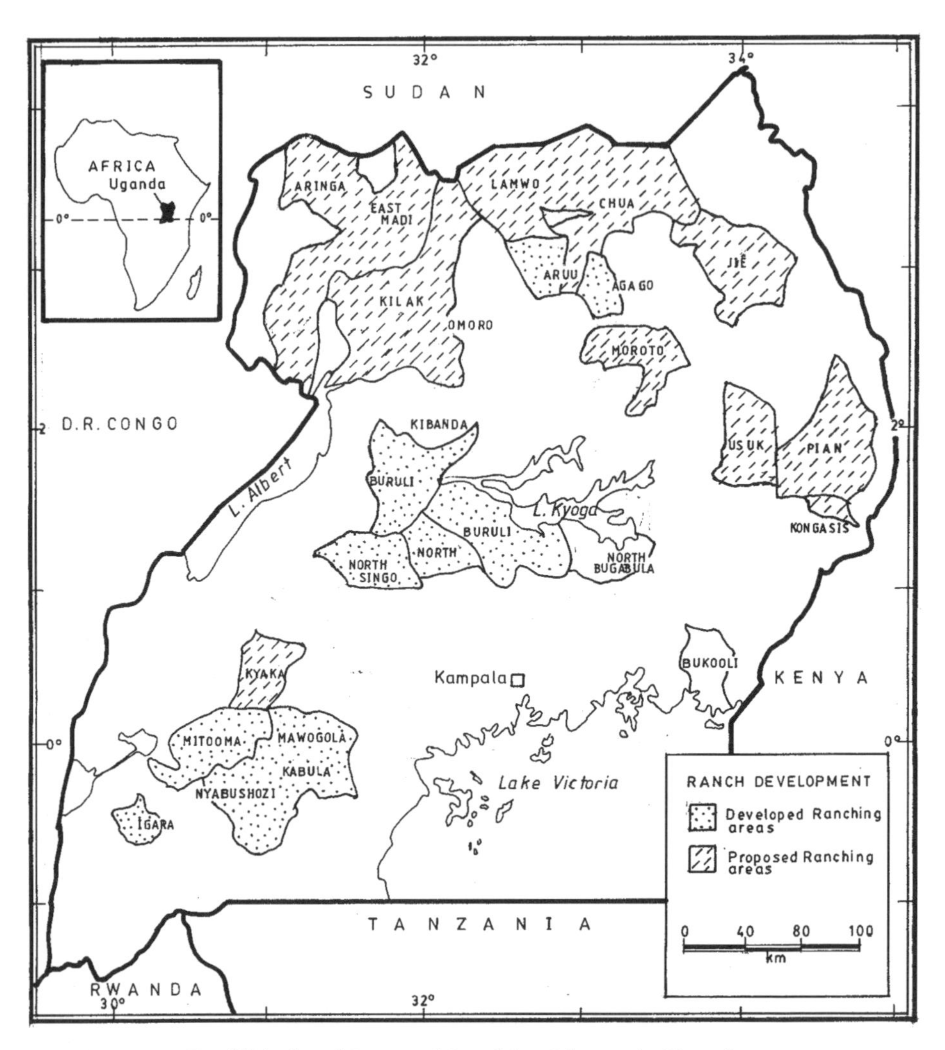

FIGURE 3 *Established and Proposed Ranching Schemes in Uganda*
Source: 2nd Beef Ranching Project, March 1975 – Statistics Section – MAAIF

2.5 The Achievements of the Ranching Schemes

In the first five years of its operation (1962-1968), the achievements of the ranching enterprise were described in the following terms:

> The past five years in Uganda have seen major developments in all aspects of the livestock industry. From small beginnings on experimental stations, commercial milk production has expanded until now there are 20,000 exotic or cross-bred dairy cows, and producers are delivering 7,000 gallons a day to 18 dairy plants. Production is expanding daily. Similar from one commercial ranch in 1962, with 400 herds of indigenous cattle for beef production, we now have over 70 ranches with 30,000 head of beef cattle, of which one third are exotic, exotic crosses or Boran. The biggest single development in ranching so far is the Ankole-Masaka Ranching scheme. With 13,000 head of cattle now, it should have 75,000 head when the proposed extensions are completed and when all the ranches are fully stocked.[65]

For phase II ranches, especially in Kabula, initial developments including perimeter fencing, valley dams and dip construction were carried out by the Uganda government and an American fund. Competent members of the public were called upon to occupy these ranches. These ranchers were given an initial probation period of five years, which they all passed and were given a full time lease of 49 years. In fact the majority performed to the satisfaction of government that had to come in some cases to regulate the number of beasts on any given ranch.

TABLE 1 *Number of cattle on the Ankole ranches between 1965 and 1970*

Date	*Estimated No. of cattle*	*The number of ranches*
December 1965	3,000 heads of cattle	14 ranches including 3 from the LUIU
December 1966	4,100 heads of cattle	18 ranches
December 1967	7,302 heads of cattle	27 ranches
December 1968	11,615 heads of cattle	35 ranches
December 1969	16,058 heads of cattle	39 ranches
March 1970	16,458 heads of cattle	39 ranches

Source: *Memo to RAC* of 14 April, 1970 by I.B. Katetegirwe on occasion of RAC visit to Ankole Ranching scheme.[66]

Because of the success of the ranching schemes established in phase I, there was a proposal to establish ranching schemes in other parts of the country during Obote 1. The idea was carried forward by Idi Amin's government. Of these proposed schemes only the Buyende scheme in Bukholi was demarcated and partitioned into 16 units of approximately 600 hectares each. However, because of the political and economic problems of the government at the time, no further development was undertaken at Buyende. The other areas that had been earmarked for ranch development included West Madi, East Madi, Zoka county in Moyo District, Jonam County in Nebbi district, Palabek in Kitgum, Bokora and Pian County in Moroto District, Singo county in Mubende and Kyaka County in Kabarole. These areas were considered agriculturally marginal and were tsetse fly infested as well.

It was anticipated that the development of ranching schemes in Pian and Bokora counties would not only act as buffer to cattle rustling, but would in the long run have a demonstration effect on nomadic Karimojong cattle keepers. These proposals were abandoned because of the subsequent difficulties faced by the Idi Amin government.

2.6 The Development of Livestock Production: The Idi Amin Era (1972-1979)

When Idi Amin grabbed power in a military coup on 25 January 1971, a solid foundation had already been laid for the ranches developed in the first phase of the Ankole and Masaka ranching schemes. In addition government ranching schemes in Singo, Buruli and Bunyoro were operational by 1974. A lot of animal breeding was taking place on the departmental ranches. Many of the ranches had mixed breeds of livestock for beef. Idi Amin, in an attempt to create an indigenous economy, had expelled Asians in 1972 and nationalised all foreign-owned business enterprises. The nationalised enterprises were allocated to political and military supporters. The ensuing economic mismanagement and political isolation grievously affected the livestock sector.

Initially, the performance of the veterinary department and of the Ministry of Land and Surveys responsible for the government ranching schemes operated according to rules. By 1975, the handling of ranches had become a disaster. Cabinet Ministers and other authorities in these ministries gave away ranches to their political supporters, and repossessed others in a most absurd manner. Ranches that had been recently established in Mawogola, Singo and Bunyoro, were allocated without any considerations for the original criteria. In 1990, the then Minister of Animal Industries and Fisheries told the National Resistance Council during the debates on the issues of ranches that:

It was enough to know the Minister of the Chairman of the Land Commission and everybody became an officer and they allocated. In fact some people who were allocated these ranches, even today (1990) they have probably never gone there. I remember one case, I think he was a big military man, he was allocated a ranch. He went to this place hoping to find the place teaming with cattle. He found empty land. He said, is this the ranch you were talking about He has never gone back again.[67]

Because of the nationalisation of foreign owned assets pursued during Amin's economic war and subsequent poor human rights record, USAID suspended its financial support to the livestock sector development. Idi Amin turned to the Arab world, and managed to secure a line of credit from the Kuwait Fund for the first five-year Beef Sector Ranching Development Project (1975-1980). However, funding from Kuwait for beef ranching development could be realised due to the liberation war that started in 1978. All the material that had been procured to develop these ranches were looted at field stations.

Ranching infrastructure was destroyed because of lack of maintenance. Livestock on the ranches began to die because of lack of disease control facilities and chemicals. For example, by October 1978, ranch number 16 in the Singo ranching scheme had 18 pure bred Charolais, 10 pure bred Boran and 1112 crosses of Charolais and Boran.[68] By 1978, a report on the status of the ranching schemes indicated the following:

> The ranch roads are in a terrible state. They need to be re-graded immediately. Most of the perimeter fences are broken down. These originally were of plain wire. I propose that the ranchers are assisted either in buying barbed wire from Casements or they buy from our stores, whichever is feasible.[69]

The peak in livestock production, in terms of cattle population was reached in 1978. As shown in table 2, livestock numbers began to decline on all different sub-sectors of livestock production. The worst affected were the departmental ranches, which reflected the crumbling of the government's administrative machinery, especially veterinary extension and input service delivery.

2.7 The Obote II Period (1980 and 1985)

2.7.1 Government's Livestock Development Policy

During its second term in power, the Obote II government launched a rehabilitation programme to revive the economy and stimulate its growth. A 10-

year action programme was drawn up in September 1981, and subsequently a recovery programme for 1982-84 was produced in April 1982. In the agricultural sector, the policy and administrative reforms introduced, targeted identified constraints. The most serious constraints to livestock production were identified as the breakdown of disease control leading to large losses of stock due to East Coast fever and other diseases, and a general lack of input and facilities due to shortages of foreign exchange. The recovery programme for the livestock sector focussed on restoring animal health control services and supporting private sector activities such as the input supply and the restoration of milk collection services. The government earmarked the rehabilitation of over 400 private and parastatal ranches (Republic of Uganda, 1984).

During the Obote II period, the major projects undertaken for the development of the livestock sector targeted a small section of the pastoral population. These projects did not benefit the majority of the cattle keeping communities; they majority were living in the war-zones. These projects included the rehabilitation of the dairy industry. This project was intended to help the diversification of agriculture by redeveloping milk production and distribution in order to meet the increased demand for milk in the urban areas, where the major of the milk supply was re-constituted. Funded by UNDP from November 1981 to the first half of 1985, the project mainly benefited dairy farmers in the outskirts of Kampala where there already was relative security.

The Animal Disease Control project, which was to cover 24 districts, was designed to increase the capacity of the veterinary department to be able to combat animal diseases more effectively. It was mainly affected by the absence of veterinary extension staff on the ground. This project was started in 1982 to1984 and financed by the European Development Fund (EDF). It was supposed to provide veterinary staff with transportation, cold storage equipment, drugs, vaccines etc. Repairs of dip tanks, spray races and cattle treatment crashes were also to be undertaken.

Through technical assistance provided by the EDF, the government was able to undertake the renovation of the artificial-breeding centre in Entebbe. The Agricultural Reconstruction Programme which covered the seven districts including Lira, Kitgum, Apac, Gulu, Soroti, Kumi and Tororo, was to provide veterinary input, and repair the animal disease control infrastructure. Perhaps the most controversial of these projects was a special one, designed for Moroto and Kotido districts of Karamoja. This project was called the Integrated Food Production and Rural Development Project, Karamoja. It was intended to resettle cattle keepers from dry areas into areas that received more reliable rainfall. This project merely increased conflicts over dry season grazing resources, which under the traditional resource management systems, would be used during the dry season only.

The government also secured a line of credit from the African Development Bank for the rehabilitation of 80 privately owned ranches in order to increase beef production. This project would also finance animal breeding in order to supply smaller ranches with cattle. The project had received 228 applications for ranch rehabilitation loans by July 1983, out of which only 49 had been approved. Of these 49 ranchers, only 30 were owner residents.

The success of this project was affected by bureaucratic delays in the processing of loan applications at the Uganda Development Bank. The actual importation of the breeding stock took a long time. This posed considerable difficulties because there were inadequate quarantine stations. The importation of large numbers of Boran breeding stocks turned out to be very costly. It became questionable whether the ranchers would be able to repay interest and capital repayment charges on the loans.

2.7.2 The Civil Wars and Livestock Production

As the ranches were beginning to pick up during the Obote I period, a civil war broke out starting at the Luwero Triangle, covering much of the cattle corridor. The various livestock production sub-sectors had been affected differently by the liberation war of 1979. During this war, a lot of cattle was stolen. Some cattle was driven to Tanzania, others were eaten by soldiers and the balance died because of the breakdown of disease control facilities. From the statistics available, departmental ranches lost up to 54 percent of their cattle in varying degrees. For example, Bunyoro lost up to 100 percent of its cattle population between 1978 and 1979, Singo 84 percent, Maruzi 79.5 percent, Bukaleba 75.6 percent, Nshara 60 percent, Ruhengeri 37 percent, Lusenke 30 percent and Acholi 8.2 percent.

The situation was not any better in the government ranching schemes. The Bunyoro ranching scheme lost up to 100 percent of its cattle population between 1978 and 1979. Singo lost up to 33.4 percent, Masaka 29 percent, Ankole 20 percent. It is only the Buruli ranching scheme, which registered a 3 percent increase in cattle population.

The protracted struggle against the Obote II government by the National Resistance Army led by Museveni started in 1980. At that time, the Singo ranching scheme had 23 ranches that had been fully developed and stocked with livestock. There were over 10,000 head of cattle on these ranches. By December 1982, the Singo ranching schemes was not operating normally because

TABLE 2 *Commercial Livestock Ranching Development in Uganda (1978-1983)*

	Location		*No.*	*Head*	*No.*	*Head*	*No.*	*Head*	*No.*	*Head*	*No.*	*Head*	*No.*	*Head*
Departmental ranches	Nshara		1	4,280	1	1,725	1	643	1	326	1	850	1	908
	Ruhengeri		1	1,885	1	1,180	1	657	1	333	1	626	1	600
	Acholi		1	2,662	1	2,444	1	1,085	1	272	1	34	1	22
	Maruzi		1	1,521	1	312	1	91	1	6	1	215	1	197
	Singo		1	1,104	1	173	1	124	1	68	-	-	-	-
	Bunyoro		1	922	1	looted	-	-	-	-	-	-	-	-
	Bukaleba		1	734	1	179	1	170	1	69	1	317	1	67
	Lusenke			33	-	40	-	38	-	33	-	-	-	-
Sub-total			7	13,141	7	6,053	6	2,808	6	1,107	5	2,042	5	1,794
Uganda Livestock Industries ranches[1]	Kiryana		1		1		1		1		1		1	
	Kyempisi		1		1		1		1		1		1	
	Aswa		1		1		1		1		1		1	
	Maruzi		1		1		1		1		1		1	
Sub-total			4	33,919	4	41,000	4	36,000	4	28,192	4	16,411	4	7,955
		Plan	Oper.											
Ranching schemes	Ankole	50	47	43,896	45	35,212	45	35,795	44	34,564	44	34,564	47	35,608
	Masaka	59	39	27,366	40	19,444	44	16,712	51	16,191	47	15,990	36[2]	17,702[2]
	Buruli	27	27	13,202	15	13,712	15	13,740	15	13,768	-	-	-	-
	Singo	34	19	11,151	15	7,421	15	9,791	15	6,500	5	3,500	-	-
	Bunyoro	37	10	4,215	-	looted	-	-	-	-	-	-	-	-
Sub-total		207	142	99,830	115	75,789	119	73,038	125	71,023	96	54,054	83	53,310
Other individual ranches (totals)			318	150,304	376	196,373	393	396,165	427	395,444	-	-	-	-
Grand total			471	297,194	502	319,215	522	508,011	562	495,766	105[3]	72,5073	92[3]	63,059[3]

1 No annual data for individual ranches available in the years 1878-1980; 2 Excludes squatted ranches and cattle thereon; 3 Excludes figures from individual ranches which were not available.

Source: Republic of Uganda (1984:70)

of the political insecurity.[70] The government ranch number 16 in the Singo ranching scheme was looted in November 1982, and then abandoned. There were just a few ranches in Singo that still had some animals on it by December 1982. These ranches included ranch number 23 with 180 herds of cattle, number 24 with 260 herds of cattle, number 1 with 289 herds of cattle, number 2A with 450 herds of cattle, number 2B with 395 herds of cattle and ranch number 4 with 410 head of cattle.[71]

By 1986 the infrastructure on all the ranches had been destroyed. A few months after the end of the war, only seven ranchers were trying to rehabilitate their ranches.[72] A report prepared for the Ministry of Animal Industry and Fisheries in 1986, divided the development in Singo ranching scheme into the following categories:

i In the first category there were ranches that were well-developed and included ranches number: 1, 2, 4, 5, 9, 12, 13, 14, 16, 17, 18, 19, 21, 23 and 24. These ranches had been stocked with cattle ranging from 400 to over 1000 herds of cattle each.

ii In the second category there were ranches number 3, 6, 7, 8, 10, 11, 20 and 22. The construction of cattle dipping tanks had been completed but these ranches were not stocked with cattle.

iii In the third category, were ranches number 15, 25, 26, 27, 28, 29, 30, 31, 32, 33 and 34. Ranch number 15 had been allocated when the perimeter fence and water valley tank had been completed but the owners never took care of it. Of ranches from number 25 to number 34 although, although they had been allocated before being demarcated, only those who had been allocated ranches number 25 and 30 had taken the initiative to have land surveyed and water valley tanks excavated manually.

On the basis of the developments in the schemes, the report recommended that those allocated ranches number 3, 6, 7, 8, 10, 11, 20, 22, 25 and 30 be given a time limit to develop them. Ranches number 15, 26, 27, 28, 31, 32, 33 and 34 were to be re-allocated. Number 29 had already been re-allocated.[73]

The perimeter fencing for Maruzi departmental ranch was done in early 1968. The fencing was financed by the government and technical assistance was provided for by the FAO/UNDP. The first batch of cattle was introduced in 1971. The cattle population on the ranch increased from 258 in 1974 to 994 in 1975, to 1,174 in 1976 and 1,330 in 1977. The ranch attained a peak cattle population of 1,521 in 1978, after which the number of cattle started to dwindle. The numbers fell to 312 in 1979, and 91 in 1980. It was recorded that in 1981, there were six heads of cattle on the ranch. In late 1982, efforts were made to restock with 215 local Zebu using a line of credit made available to the government by the African Development Bank (ADB). The efforts were short-lived. All heads of cattle were looted in late 1983. Soldiers first occupied the ranch in 1983 after which the ranch was occupied by different successive groups of soldiers until 1986 when the ranch was again looted on a larger scale.[74]

The declining trends in the livestock production sector is captured more vividly by the figures on cattle population numbers in table 2.

2.7.3 Constraints to Livestock Production

The political instability resulting from the civil wars that started with the 1979 Liberation War and lasted until 1985 led to a shortage of foreign exchange. In turn the resulting poor economic performance led to severe deterioration in disease control, upon which government policy for livestock improvement had been anchored. As identified in a review of the beef ranching sector (Republic of Uganda, 1984), the problems of livestock productions associated with these difficulties included the following:

i On the ranches where the breeds included the exotic and the crosses, there were increases in resurgence of East Coast Fever and Trypanosomiasis because of a severe shortage of drugs and acaricides. This led to significant losses of stock.

ii There was a re-infestation of tsetse flies in the belt between Kiryana/ Kyempisi-Bunyoro-Maruzi ranches.

iii The majority of the ranches were in areas that were considered insecure. Many of the veterinary extension staff were not at their stations. Those who were had a very low morale because of the poor facilities and the general rundown of support services.

iv The implementation and monitoring of disease control was greatly hampered by the breakdown in communication services including the road networks and telecommunication.

The civil wars affected the livestock industry in general. The ranch owners in the government schemes faced the same conditions, as did individuals who had private ranches outside the government ranching schemes.

However, owners of private ranches were better able to take the initiative in recovering from the effects of the war, than owners of ranches in the government ranching schemes. The former were mainly individuals who relied more on the incomes obtained from their ranching enterprises, and were resident operators of the ranches which they had developed using their own resources. The survival of these private ranch owners depended on the functioning of their ranches. No private ranch owners were found in the areas around the Bunyoro ranching scheme. Buruli and Ankole ranching schemes had several private ranches near the government schemes. The resident ranch owners had more incentives to safeguard their investments, than did either those who had alternative sources of income or those that had acquired huge loans to develop the ranches in the government schemes. Many of the resident owners did not abandon their ranches.

2.8 Squatter Pastoralism on the Ranching Schemes

2.8.1 The Origins of Squatter Pastoralism

In the preceding discussions, we had already intimated that even before the establishment of the ranching schemes, there was the problem of landless squatter pastoralists, for example in the Ankole Kingdom, and that the problem of squatters was as old as the ranching schemes.

Many of the traditional cattle keepers who were not allocated a ranch during the 1960's lived on pieces of land adjacent to the ranching schemes, especially the communal grazing areas. In Ankole, some areas were never included in the land to be earmarked for the ranches because these areas contained many tenants with cultivated areas. This would have implied paying these tenants heavy compensation. Once these people who remained on land in between ranches began to acquire animals, they became squatters on the adjacent ranches. The problem of squatters on the ranching schemes had become evident as early as the beginning of the 1970's. A delegation of ranchers from Buruli ranching scheme visited the Minister of Animal Industry, Game and Fisheries in April 1971 and complained about the following:

i Cattle from outside the scheme were grazing on their ranches
ii There were a lot of cattle thefts
iii People outside the scheme were burning grazing pastures indiscriminately

In May, 1971, in a meeting between the Buruli ranchers and the Regional Veterinary Officer, Buganda, the ranchers revealed the following problems:[75]

i There were pockets of mailo land within the ranching scheme. Cattle were kept on these pieces of mailo land. Cattle owners would cut fence wires so that they could graze on the ranches. The department was asked to purchase all the mailo land within the ranching scheme and allocate it to the ranches. After all the pieces of mailo land had been purchased it would be possible to control the spread of cattle diseases and cattle thefts.

ii When the defunct Buganda government undertook the development of the Buruli ranches, it compensated the people in these areas. However, some were still living on the ranches: ranch 5B (2 tenants), ranch 5C (2 tenants), ranch 3A (1 tenants) and ranch 3B (7 tenants). These tenants were practising subsistence agriculture while others kept cattle. The presence of the squatters on these ranches was considered a hindrance to the development of the ranches. The department was requested to compensate the squatters so that they would leave the ranches.
iii Livestock owners who were living adjacent to the ranching scheme did not have enough water for their livestock. They tended to water their animals in the valley tanks within the ranching scheme, cutting the fence wires and spreading diseases. The department was asked to construct valley tanks for the livestock owners residing near the ranches.
iv Some of the ranchers allowed cattle owners to graze cattle on their ranches, collecting an annual fee from them. Such squatters sometimes indulged in cattle thieving. The department was asked to carry out checks for squatters frequently so that those ranchers who were collecting fees would be dealt with.

In September 1971, the Regional Veterinary Officer, Buganda, in a brief to the RAC, identified squatting pastoralists as a major problem that affected Buruli ranching scheme, and were the result of the following:[76]
i Some people who had settled on some ranches complained that they were never compensated by the Buganda government, although past records showed that there were squatters only on one ranch (ranch number 6A), and not on any other.
ii There were pockets of mailo land which were never purchased by the Buganda government when the ranching schemes were established. Squatters settled on such pieces of land.
iii Only a few ranches were managed as commercial enterprises.

The problem of absentee ranchers had already been made note of by 1971. Many of these ranches had very poor ranch managers. Some of the ranchers invited herdsmen to graze their cattle within the ranches. They cared little about proper ranch development. In its meeting of September 1971, the Ranching Advisory Committee noted the following on the problem of squatters in Buruli:

> Squatters in the Buruli schemes (as opposed to mailo landowners) are mostly nomadic and the problem of trespass could be largely solved by erection of perimeter fencing, and the provision of water outside the

> scheme. The Government valuer has been asked to assess and negotiate for Mailoland in the scheme.[77]

Requests were made by Commissioner of Veterinary Services to assess and compensate Buruli squatters in September 1971. The Secretary to the Treasury replied:

> Your request for assessment and compensation of property of squatters and the mailo land owners has been considered carefully, but in view of the present financial difficulties, I would suggest that this exercise be deferred until the following financial year.[78]

In Mawogola areas, 131 tenants were compensated to a total value assessed at Uganda shillings 1,033,736.05. The highest amount received was shs. 275,576.90 in Kitayira village, Soga Parish, Mijwala sub-county Mawogola county, and the lowest amount received was shs. 57.50 in Kyamamba village, Byakajula Parish, Kahiro sub-county, Kabula county.[79] At the Singo ranches, there was a similar problem of tenants squatting on some of the ranches, which the ranch owners claimed made it difficult for them to develop their ranches.[80] A request to assess their property for compensation was granted to the Commissioner of Veterinary services, and the squatters were given six months notice to quit.[81]

The establishment of the ranching schemes attracted private developers to lease out large chunks of land in the areas around the government ranching schemes, further alienating the traditional nomadic cattle keepers. At one time, the Secretary of the Ankole RSB withdrew an objection to grant a lease of 202 hectares of land adjacent to the Ankole ranching scheme. This land had been offered to one Mrs. Zerida Rwamuranga in 1971, on the condition that the lease offer include clauses to protect stock in the development area.[82] These private developers, like their counterparts in the government ranching schemes, began to fence off their land to ward off landless cattle keepers. The particular developer above, complained later to the Ranches Selection Board that on beginning development of the 202 hectares, she had been subjected to harassment by nomadic cattle keepers who kept moving their cattle into the area from the North of the Mbarara-Lyantonde Road. The movement of cattle by the nomads into the area where the ranches had been established was organised by a local chief and because it was within the same sub-county, the Ranches Selection Board could not prevent the movement of the cattle. The nomadic cattle keepers and their cattle were therefore seen as presenting a threat to the Nshara government ranch and to cattle in the Kachera government ranching scheme. It was proposed that all settlers in the area be compensated and moved out so that the area would be incorporated in ranch development.[83]

During Amin's government in Ankole, ranch owners invited squatters onto their ranches in order to meet the livestock returns that would enable them keep their ranches, which was then not the biggest issue for the RAC, as long other recommended husbandry practices were implemented. The Secretary of the RAC noted:

> I refer to the attached minutes of the RAC meeting of 31 January, 1972. When I was in Ankole last week, I discussed ranches with Mr. Katetagirwe, SAHO, Sanga. On ranch 27, there are now 490 herds of cattle. No doubt many of these cattle do not belong to Mr. Bitembe but I think it will not be possible to prove that point. The management while not being particularly good is not outstandingly bad on any definable point, and effective tick control is being carried out. As you are aware the Committee felt that Mr. Bitembe would never be a really good rancher, but I have considerable doubt if the case for eviction can be shown to be clearly justified.[84]

There was a very serious crisis of squatters at the Singo ranching scheme. A report on the Singo ranching schemes indicated that squatters in Singo ranching would not have been a problem if all those who had been allocated ranches had taken an interest in their ranches and developed them. Badly developed ranches attracted squatters who even let their cattle graze at developed ranches.[85]

Some cattle keepers, after putting their cattle at departmental ranches, used other's absence of incentives to strengthen their applications to claim a ranch. Take the following two examples: Bihogo ranchers unsuccessfully applied for a ranch in 1970. In 1971, the Director of Bihogo ranchers wrote to the Secretary of the RAC that:

> We want to register our great concern for the application for a ranch, which unfortunately has so far been unsuccessful. We are presently on a government ranch, and it is most unsatisfactory. We are unable to develop, and cannot buy because the government continues to stock it as and when the number of animals decreases. We are also unable to invest any more money in the venture because of uncertainty in which we are operating. We cannot get any credit to develop the venture because no banker will give us any credit unless we have a base, i.e. some kind of title for the land on which we graze the animals.[86]

In 1972 he wrote to the Secretary of the RAC, that he was squatting, together with others on Government Reserve Ranch number 25, and complained:

> Fellow squatters kept on bringing in new animals; they could bring in other diseases, there is no way of checking or stopping such a risk.[87]

During the 1980's, some cattle owners became squatters of their own accord, for instance by negotiating with the veterinary department to be allowed to graze the animals. A cattle owner in Buruli applied to allow his cattle to graze on ranch number 4A, Kalyampande, because he had been sharing with four other parties on another ranch where grass had become very sparse for all of them.[88]

2.8.2 The Alienation of Critical Dry Season Grazing Areas

It is perhaps the 1979-liberation war that brought in a new wave of cattle keepers displaced from areas bordering northern Tanzania. Most of these cattle keepers settled in the Lake Mburo area. The area, now the LMNP, was initially controlled hunting ground under the Ankole Kingdom. It was first gazetted as a game reserve in 1964 and since then had experienced a high rate of population growth. From the original 115 families granted resident permits in 1964 because they occupied parts of the area, the number of families in the area grew due to political instability and corruption, to 363 by 1977. Between 1971 and 1979 government officials, churches and individuals went into a land grabbing spree and occupied 50 square miles of the reserve, further reducing the amount of resources available to pastoral groups.[89]

The other category of squatter cattle keepers in the Lake Mburo area were situated on the Nakivale Rwandese Refugees Resettlement scheme, located south of Lake Mburo. The majority of the Rwandese cattle keepers were mainly Tutsi's who had come during the 1950's probably as political refugees. There were also a few Hutu's who had come as economic refugees in search of wage labour on coffee plantations in different parts Buganda. Many of them became farm labourers in Ankole and Masaka and inter-married with the local population. Having had a pastoralist background, they were more predisposed to livestock. Many of them even used their wages to acquire livestock. Since they did not own land, they ended up being nomads. Survival of their herds was all they looked to. They would move their cattle wherever it was considered favourable for their survival and multiplication. They traversed the entire savannah grasslands across much of Uganda as if it was one big grazing area.

In 1980, the now defunct National Wildlife Committee (NWC) recommended that the Lake Mburo Game reserve be turned into a National Park.

Following this recommendation, two senior government ministers, one of them the minister responsible for wildlife and the other holding the agriculture portfolio and also local area member of parliament decided that part of the area, the Rwanjeru strip, be privatised and shared among themselves. As they could not agree on who was to take which part, the Agriculture Minister decided to put livestock on the best part. Unable to accept this the Wildlife Minister, using the 1980 NWC recommendations, declared the whole area into a National Park in 1982 and therefore a protected wildlife conservation area (GAT-Consult and Norconsult, 1992).

The manner in which the decision to turn the game reserve into a national park was implemented during the Obote II government, showed that the matter had been overly politicised. The local inhabitants of the areas were never consulted about government plans for the area. The immediate cause for the establishment of the park was not motivated by a desire to achieve social justice or even improve the social welfare conditions of the cattle keepers inhabiting the area, but were instead selfish economic, and political motives by leading government officials.[90] The violent manner in which the cattle keepers were evicted amounted to a gross violation of human rights.[91] The eviction of these pastoralists from the park, left massive violations of human rights in its wake including actual death, torture, damage and destruction to property and great human suffering.

Many of the Rwandese pastoralists were initially settled in the Oruchinga valley refugee settlement camp, south of lake Mburo. Over the years, many moved out of the refugee settlement camp and were integrated in the local communities. Those who had cattle, moved out in search of grazing lands. The majority ended up inside the then Lake Mburo Game Reserve. When it was upgraded into a national park in 1982, the pastoralists in this area were forcefully evicted, and attempts were made to re-confine them within the resettlement scheme. The majority of the cattle keeping Rwandese that had tasted the freedom outside refugee camps dispersed to other cattle keeping areas.

Almost all of them ended up first in Ankole ranching scheme, but later dispersed further northwards into Mawogola and Kabula ranches of the Masaka ranching schemes, and into Singo and Buruli ranches where many ended up in then-NRA guerrilla controlled areas. Some of the Rwandese pastoralists had become naturalised citizens of Uganda, and also came to be referred to as Banyarwanda of Uganda. Many came during the coup of 1959 and were employed as herdsmen by cattle keepers in the ranching schemes. Over the years, as they acquired animals, many opted to move out on their own. They swelled the numbers of nomadic groups that roamed the Savannah stretch of Uganda. They were the main category of cattle keepers from the different ranching schemes who were, without exception, completely landless. As they accumulated cattle over the years, they became a formidable pressure group when the restructuring exercise had started.

Among the displaced indigenous cattle keepers, there had always been a persistent disgruntlement against loosing their traditional grazing when the ranching schemes were created. After the establishment of the ranching schemes the people became squatters wherever they went. The ranching schemes caused the traditional nomads to be landless. For other cattle keepers, periods of scarcity caused by drought were usually very difficult times. They saw their plight as a direct result of their failure to gain access natural pastures and water points either enclosed in national parks or on individual ranches. Successive governments in the 1970's and 1980's did not take any steps to institute appropriate land use policies. The cattle keepers could only gain access to resources by forcing their way in.[92]

2.9 Livestock Development under the National Resistance Movement (NRM)

2.9.1 NRM's Policy on Livestock Sector Development

After assuming power in January 1986, the NRM government formulated an economic recovery programme focussing among other issues on improving and transforming the existing livestock production systems. In this undertaking the pastoral communities were involved. The pastoral nomads were addressed by and became involved in the following projects. Disease control, the improvement of diagnostic facilities, intensified veterinary extension and private veterinary practice, the training of community animal health workers, supporting private sector participation in the marketing of livestock and their products and in the provision of water and other infrastructure address these pastoral nomads (MAAIF, 1996). After 1996, the NRM government came up with an agricultural modernisation agenda, as a fulfilment of President Museveni's 1996 presidential campaign pledge.[93]

In the agricultural modernisation strategy (1996-2000), the government has undertaken to promote the specialisation of agricultural activities based on agro-ecological zones (MAAIF, 1996). The main objectives of the strategies adopted for the livestock sub-sector in the modernisation agenda, include reversing the decline in livestock numbers by improving disease control, and adopting other productivity enhancing measures such as increasing the availability of water for livestock by constructing water harvesting facilities (Republic of Uganda, 1997). One of the main strategies adopted by the NRM has been the settlement of nomadic cattle keepers. From the point of view of the government, the strategy of settling pastoralists is considered as a panacea to resource management problems associated with nomadic practices which the NRM considers to be among the leading obstacles to the development of the livestock production sector in Uganda.

A livestock census carried out on the ranches in 1990 indicated that of the 50 ranches in the Ankole ranching scheme, 4 ranches (8 percent) did not have cattle on them. It was on 17 ranches (34 percent) only, that ranch owners had more cattle than squatters. On as many as 29 ranches (58 percent), the ranch owners had fewer cattle than squatters. In Masaka ranching scheme, out of the 59 ranches, there were no cattle on 29 ranches (49 percent), and squatters had more cattle than ranchers did on 25 ranches (42 percent). Out of the 34 ranches in the Singo ranching scheme, squatters had more cattle than ranch owners did on 22 ranches (64.7 percent). Out of 27 ranches in Buruli, squatters had more cattle on 19 ranches (70 percent). In Bunyoro, out of 37 ranches, 29 (78 percent) did not have cattle on them. Squatters had more cattle than ranchers did on 7 ranches (18.9 percent). These figures are summarised in table 3.

TABLE 3 *Cattle owned by ranchers compared to squatters in 1990*

		Number of ranches on which:		
Ranching scheme	*Number of ranches*	*No cattle exists on the ranches*	*More cattle owned by ranchers*	*More cattle owned by squatters*
Ankole	50	4 (8%)	17 (34%)	29 (58%)
Masaka	59	29 (49%)	5 (8%)	25 (42%)
Singo	34	5 (14.7%)	7 (20.5%)	22 (64.7%)
Buruli	27	1 (3%)	7 (26%)	19 (70%)
Bunyoro	37	29 (78%)	1 (2.7%)	7 (18.9%)
Total	207	68 (32.8%)	37 (17.8%)	102 (49.2%)

Source: Ranches Restructuring Board (1997)

The government's concern and the need to settle landless cattle keepers who were squatting on ranches in and outside the government ranching schemes was justified. In 1990 the Minister of Agriculture Animal Industry and Fisheries, Prof. Mondo Kagonyera, argued that the contribution of squatters to the country's economy was not insignificant. He said:

> There is nothing wrong with the government being concerned for social or economic reasons and even legal reasons. Because these people could chal-

lenge any government at any one time for having abused their human rights![94]

By 1998, the contributions of the squatters were no less insignificant, as they owned a total of 60,000 heads of cattle. Two major policy interventions by the NRM to settle the pastoralists were the establishment of the Kanyaryeru resettlement scheme (following the degazetting of part of the Lake Mburo National Park) and the restructuring of the former government sponsored ranching scheme. Both interventions made land available for the settlement of formerly landless cattle keepers. The government strongly believes that settling pastoralists would provide the required incentives for hitherto landless cattle keepers to invest in the improvement of land that they own. This leads to the adoption of improved pastoral resource management practices.

2.9.2 The Resettlement of Landless Cattle Keepers

When the NRM came to power in 1986, one of daunting tasks it was faced with was the resettlement of displaced people, including the cattle keepers who had been dispossessed when the ranching schemes started in the 1960's. These pastoral nomads continued roaming the entire stretch of the dry grassland Savannah from Ankole, through Masaka, Kabula into Singo through Bunyoro and stretching eastwards into Teso-Karamoja.

Pressure mounted from different political pressure groups for the new government to recognise the pastoralists contributions by accenting their demands. The cattle keepers exerted such a source of pressure on the NRM. First and foremost, not only cattle keepers like Museveni was, but also 'Banyarwanda' (this term was politically used to describe the majority of western cattle keeping tribes that did not support Obote's government) as Museveni was referred to, was enough to earn the wrath of the Obote II government. Faced with harassment, the majority of cattle keepers joined the NRA in the bush for their survival, sent their sons and daughters to fight the government or sacrificed their property to support the bush war from wherever they had taken refuge. After NRM captured power in 1986, they presented a very formidable pressure group on the government. One of the key demands from the cattle keepers was to be resettled in the ranching schemes. Those who had been displaced from the areas where ranches were established, especially the areas where ranches were set up in the late 1970's such as Singo and Bunyoro ranching schemes, were demanding to be resettled in the areas from where they had been displaced.

Immediately on assuming power in 1986, the NRM set up the Lake Mburo Task Force, whose terms of reference were to iron out the issue of settlement in the Lake Mburo National Park (LMNP). The area around Lake Mburo is endowed with a unique ecosystem, which provides conditions capable of supporting a variety of habitat types, unmatched in many places of the World. It provides a habitat to a multitude of fauna (both large and small wildlife stock) and in the dry season it is the only area that retains permanent water around Lakes Mburo and Kachera. There is a lot of contestation for both pastures and water between cattle keepers and their livestock on the one hand, and the Park's management and their wildlife on the other. The magnitude of such contestation increased in the 1970's and 1980's mainly because of a population explosion of cattle keepers who began to live with their livestock near the boundaries of the park, and as time went on, moved inside the Park boundaries. At the same time, the number of pastoralists who encroached on the Park also increased, especially as poorly maintained water sources in the government ranching schemes began to dry up sooner rather than later during dry seasons.

Pastoral encroachment on LMNP was and still remains the single most significant problem that threatens the survival of the Park. This was demonstrated by almost every study carried out in the area (e.g., GAT-Consult and Norconsult AB, 1992; Kamugisha and Stahl, 1993; Busenene, 1993). Different studies focussed on different levels of encroachment. The numbers of encroachers varied seasonally, and was usually higher during the dry seasons when there is a scarcity of pastures and water in all other areas. In 1992, a study showed that up to 70 percent of the grazing lands had been left bare by overgrazing encroachers' cattle. 300 families were residing in 100 square miles of the Park, out of which 200 were pastoralists with 20,000 herds of cattle.[95]

The Lake Mburo Task Force recommended in 1987 that 150 square miles be chopped off from the Park for ranching, cultivation and resettlement of landless cattle keepers displaced from the Luwero Triangle, leaving the Park with 100 square miles, less than half of its original size. This led to the creation of the Kanyaryeru Resettlement Scheme. However, the creation of this resettlement scheme did not ease the pressure on the LMNP overgrazing and water supply, especially during the dry season. This was because as recently as 1991, private ranches occupied 50 square miles of prime grazing lands in the lake Mburo ecosystem. The Nshara Government Ranch occupied 30 square miles, the Kanyaryeru Settlement Scheme 30 square miles, and 40 square miles was occupied by private ranches and already existing agricultural settlements. The Kanyaryeru Resettlement Scheme only provided a limited respite for grazing pressures, especially for cattle keepers who remained largely excluded from the larger areas under government control and private ranches.

In the meantime, pressures on the NRM government were mounting in not only the Ankole ranching scheme, but also in the other four ranching schemes of Bunyoro, Buruli, Masaka/Mawogola and Singo. Around the time that the NRM was busy planning the resettlement of landless cattle keepers in the degazetted section of the LMNP, an organised resistance against rancher-landlords by landless squatters in the Bunyoro Ranching Scheme was reported in November 1986. This resistance followed an attempt by the NRA to evict some squatters from a ranch in order to start an NRA ranch in the scheme.[96] The squatters blocked the road and refused to allow an army truck to proceed to its final destination. Attempts by the army to occupy a ranch heavily settled by squatters were successfully resisted. It was then that it was announced that the government was in the process of reviewing all government ranching schemes. In the meantime the squatters were told to stay wherever they were.

The resettlement of landless cattle keepers in the Kanyaryeru resettlement scheme increased pressures from other cattle keepers on the LMNP and the other ranches in the government sponsored ranching scheme, as well as private ranches in the area. Even threats by the government to prosecute nomadic pastoralists did not succeeded in dissuading them from encroaching on the Park. It was at that time that the government was forced to a find a long term solution for the problem of landless cattle keepers by way of reviewing its policy towards the ranching schemes started in the 1960's. To ease the pressures on the now greatly reduced LMNP, the government made promises to cattle keepers to resettle them elsewhere, but outside the LMNP and the Kanyaryeru resettlement scheme. The government started making appeals to cattle keepers who had been living inside the LMNP boundaries to accept compensation and resettle in government ranching schemes elsewhere, mainly in Singo.

In April 1994, a total of 62 million shillings was earmarked for the Lake Mburo Conservation Project to compensate 179 squatters inside the LMNP, who would be resettled in Singo county, Mubende District. Although no squatter was forced to take compensation money, those who had no alternative place to move to, or in anticipation of land allocations in the restructured ranches were not permitted to extend their developments on land inside the national Park. It was explicitly stated that the compensated parties were expected to renounce any claims to the resources in the Park after receiving compensation and payment in full.[97] The then Mbarara District Central Government Representative (CGR), was reported having made an appeal to 200 families encroaching on the LMNP to accept the government decision to settle them in Singo. The squatters were reported to have refused to move without adequate compensation, since they had permanent assets along the edges of the Park, whose boundaries were opened in 1991.[98]

2.9.3 The Commission of Inquiry into the Government Ranching Schemes

By the end of its first year in power, it was apparent to the NRM government that the government's policy on the ranching schemes was not clear anymore. In light of increased pressures on the government ranches by squatters, the failure of many of the ranch owners to fulfil development conditions contingent to the lease offers, and political and economic problems arising from two decades of civil wars the policy needed clarification. In the beginning of 1987, the government set up a nine-man commission of inquiry to inquire into the set-up, management and work methods of the 207 ranches in the five government sponsored ranching schemes of Ankole, Buruli, Masaka, Singo and Bunyoro. The commission wanted to affect reforms and improve the efficiency of existing and future ranches. It was designated by the government under legal notice number 5 of 1987 section 2 of the Commission of Inquiry Act (Cap. 56).

In its report published in December 1988, the Commission recommended among others things, that the ranches be individually scrutinised in order to find land to settle the squatters. The Commission recommended that the government repossess 53 of the 96 ranches from individuals, companies and co-operative societies in Ankole, Masaka, Buruli, Singo and Bunyoro that had performed badly. These ranches would provide about 55,200 hectares of land on which the squatters should be settled (Republic of Uganda, 1988). The owners of these 53 ranches had become telephone ranchers, who owned large pieces of undeveloped land. Some of them had acquired the ranches fraudulently. A total of 43 ranches were identified as showing a minimum of development but due to genuine problems of instability had not been able to do much. These ranches were to be given 12 months to show concrete development. 13 of these ranches were in Masaka.[99]

The Cabinet discussed the commission of inquiry report no fewer than six times, and adopted it with slight modifications. The government's decision was based on a white paper written by the Ministry, advising the government to adopt the Commission Report's recommendations. The report sparked off wide-ranging debates with implications on the future of the ranching schemes in general and the relationship between ranchers and squatters in particular, especially the ranches with squatters.

2.9.4 The Squatters' Uprising

The Causes and Nature of the Squatters' Uprisings

The release of the report of the commission of inquiry is associated with the outbreak of violence in the ranching schemes because of the following:
a The publication of the Commission's report led to a lot of speculation from both squatters and ranchers. Recommendations about the government repossessing more ranches in Masaka than Ankole ranching scheme led to rumours to the effect that the government had acted on sectarian grounds, by repossessing ranches in Buganda and leaving those in Ankole.
b There was also speculation from the squatters that they were about to be allocated land. This precipitated an influx of squatters from neighbouring districts as far as the border with Tanzania to Masaka intensifying overcrowding in the area.[100]

Accounts by ranch owners and cattle keepers in Mawogola, Singo and Buruli areas indicated that by the end of July 1990, tension was already high between ranchers and squatters. The first clashes were reported on August 5, 1990 when squatters armed with guns, spears and machetes stormed a number of ranches in the Mawogola areas and seized them by force. By August 13, Lyatonde Police Station had reported 12 cases of clashes on the ranches in the Lyantonde areas, where all the cows, goats, sheep and buildings had been destroyed. Squatters invaded ranches, claiming that they had the right to do so. Other claimed that they had been told to do so. One was heard to remark: 'if your cattle has no grazing grass or water, and you are starving, should you wait to ask?'[101]

The squatter-rancher relationships had worsened in a short period of time mainly due to a statement by the Permanent Secretary, Lands and Survey and a Joint District Administrator (DA) Masaka/President's Office Task Force. This team toured Masaka Ranching Scheme to investigate the squatters' problems. The squatters interpreted the tour as a license to stay on the ranches on which they were squatting at the time. Ranchers tried to deny the squatters water for their animals. At one time a rancher opened fire at a squatter. Ranchers were blamed for having sparked off the scuffle by arming themselves and shooting at the squatters.

Attempts were made by government officials to try and diffuse these tensions, to redeem their credibility as leaders from these areas, and in a way to show how they were capable of handling crises. Among other people who made initiatives were members of the National Resistance Council (NRC), Cabinet Ministers and Senior Army officers who hailed from the areas gripped by the tensions. Some of these government officials played down the gravity of the situation. On August 7, 1990, an announcement was carried in the state-owned New Vision that the then Minister of State for Defence, Major General David Tinyefuza, himself a rancher in Mawogola, would hold rallies and meetings starting on August 8, in sub-district with ranchers and civil servants in the area. The meeting was intended to clear the allegations of tribalism and mounting resentment on the part of all those concerned.

Many families living a nomadic life in Masaka and Ankole had been displaced with their herds of cattle. During the protracted war of the NRA against the past government, some of these families had contributed in one way or another in helping the NRA in its armed struggles. Following the NRA victory, these displaced people began to disagree with the indigenous people living in Luwero, Bulemezi, and Kiboga areas over land issues. These families had to leave their areas slowly and others found themselves in Ankole and Masaka areas where more trouble was brewing.[102]

While the conflict was seen as one between Baganda ranchers and Bahiima herdsmen, this scenario was rejected by the then Minister of Information and Broadcasting, Mr. Kintu Musoke, who said that out of 40 ranches in Mawogola, Baganda owned only 10 ranches. And yet the violence had affected all of them without exception, the only difference was the magnitude of violence. This conflict was also widely considered as a struggle that was overtly supported by NRA soldiers who, during the armed struggle to topple Obote's regime, had been forced into joining the NRA. Those in the NRA were giving support to and backing their people (the squatters) to seize the ranches, which they claimed were established on their grazing lands way back in 1964 by the government.[103] There were reports that the herdsmen were aided, assisted and armed by their sons or relatives in the NRA, to attack and seize by force the ranches owned lawfully by large-scale ranchers who had occupied them since the 1960's. On 8 August, 1990, the Resistance Councils, Police, chiefs and courts in the area were instructed by the Commanding Officer of the 47th Battalion based at not to intervene in the ongoing 'war' between the herdsmen and the ranchers.[104]

On August 14, the government intervened and deployed the NRA soldiers to stop the destruction of property and ruthless slaughter of animals on the ranches by a mob of armed herdsmen described as squatters. At least three lorries of NRA soldiers were sent to the area to try to restore law and order. According to the Police at Police Post, six ranches were seriously affected by the wanton destruction carried out by armed herdsmen around that time. Property on these ranches was virtually destroyed and in some instances houses were set on fire. Even army units that were deployed – starting on August 14 – by the government to stop the destruction of property and ruthless slaughter of animals on the ranches, never remained impartial. They took sides with the squatters.[105] The ADA, and all police and magistrates in and Kabula received a memo from Captain Yosam Lwanyaga, the commanding Officer, which read as follows:

> You are informed to stop interfering in matters concerning squatters and ranchers. Any case that is between the above mentioned groups should not be entertained or registered by your office. The government is well aware of these problems, and it is only up to the government herself to solve them.[106]

When the clashes broke out, there were numerous attempts by ranchers to diffuse the situation, although to their own advantage. At the beginning of the first week of August 1990, Mawogola ranchers prepared a memorandum which they wanted to present to the President, outlining their grievances mainly related to the problem of lack of water and the influx of squatters looking for water and land. On August 5, 1990, President Museveni met with members of MALIFA at State House and discussed with them strategies that could be adopted as the best means for animal production in the country.

On August 8, 1990, another delegation of ranchers from Masaka and Kabula appealed to President Museveni to rescind government proposals to subdivide ranches in order to accommodate landless squatters. They appealed to the President to let the current holders of the 5 square mile ranches continue and assured him they would not complain if they were kicked out for failure to fulfil the ranch convenant they signed on allocation. They also appealed to him to step in quickly and settle the problem before the squatter question became explosive.

The government seized the opportunity. In a response to their demands, President Museveni told the ranchers that with diligent and frugal land-use, they would not need as much as five square miles to make a profitable ranching business.[107] The President went ahead to issue new measures and directives. On August 14, 1990, President Museveni, while addressing squatters at Nyakashashara sub-county headquarters, announced the measures taken by the government as preliminary steps aimed at the eventual solution of the rancher-squatter problem. He directed that cattle keepers who were not genuine squatters be given a period of one month, effective from August 13, 1990 to leave the ranches and game parks. The two groups affected by this directive were those who in the last twenty years had transferred their excessive herds of cattle from wherever they were onto the ranches. It also included those who had sold their pieces of land and opted to take their families and property to areas that covered the ranching scheme in Nyabushozi, Mawogola and Kabulasoke. These people were to go back to their former places, or buy other pieces of land. The President asked the District Administrator Mbarara to ensure that this directive of a one month notice to quit was complied with within his district. This would leave the government only with the problem of resettling genuine squatters who, in the last twenty years or so, had never owned any piece of land but had been living nomadic lives.[108]

On August 10, 1990, a three man ministerial committee appointed by the President had made a fact-finding tour of Lwemiyaga and Mawogola counties of sub-district, amid due to reports that conflicts between ranchers and squatters in the areas had become explosive. The committee comprised the Minister of Information and Broadcasting, Kintu Musoke as team leader, the Minister of Agriculture, Animal Industry and Fisheries, Prof. Mondo Kagonyera, Major General David Tinyefuza (Minister of State for Defence) and the Minister of Lands and Survey, Ben Okello Luwum. 'The atmosphere at on Wednesday morning was tense and expectant.' Although it was claimed that the situation had calmed down, the mutual resentment between ranch workers and squatters was high. Groups of either faction were seen milling about in front of shops and stalls, discussing the issue.

On August 11, 1990, the President had a meeting with the ranchers from Mbarara together with their NRC members. A brainstorming session on the issue was held. The President was informed at this meeting that a lot of property had been destroyed in Kabula. One of the ranchers claimed that a petrol bomb had destroyed his house. The President got concerned. He sent his own people to get the facts regarding this incident. But it was discovered, among other things, that the house was never destroyed. But there had indeed been some property destruction. It was decided that the ministerial committee should make a second visit to the troubled ranches in the ranching schemes.[109]

The second visit by the ministerial committee was made on 15 August, 1990. Ranches were visited in Kabula and Mawogola where it was established that squatters on the ranches had destroyed some property. The patterns of destruction were uniform on all the ranches that were visited. It was mainly installations that were destroyed. Windows were broken if these were made of glass. If the windows were wooden, they had been forcibly pulled out, and the doors broken. The easily movable property inside like mattresses, chair cushions, and other furniture were taken outside and burnt. None of the property was physically taken away. There was no evidence that any animals had been killed or taken away. There was no loss of life, although some people had scars indicating some acts of violence. Some of the squatters, many of whom still had a working relationship with their landlords (ranch owners), would come and advise them to escape when they saw hooligans coming. Most of the time when the squatters came, nobody would be there.[110]

The ministerial fact-finding committee discovered that some ranchers had decided to fence off their ranches following the government decision to subdivide and reallocate the ranches. The ranchers were speculating that the government was planning to reallocate pieces of land on which squatters were settled. They started chasing away the squatters who put up resistance. Some of the ranchers erected fences denying squatters any access to water during the dry season of 1989/90. Squatters actually cut the fences to gain access to water, to the indignation of the ranchers who started arming themselves. While some took it upon themselves to make subdivisions, and preferably to include for obvious reasons, the watering points in their own parts of the ranch, others had hired security guards and deployed them on the ranches.[111]

In fact, the army men had been sent to those areas to counter a campaign of insurgency by the ranch owners. When the ranchers learnt that the government was moving in to sub-divide some ranches in Mawogola and Kabula, they organised a campaign of terror to get rid of squatters. They went round telling the squatters that those areas were to become war zones. This was meant to create a stampede so that the squatters could flee the area. The move was to convince the government that there were no squatters and therefore, no need to subdivide the ranches in these areas.

President Museveni talked of a group of people who had organised terrorist activities with political motives in the area and thereby opposed the NRM. The gang was said to have killed five squatters on a market day, robbed a bank at Butenga and disarmed a policeman at Mpugwe. Tinyefuza said that some leaflets had been distributed in the areas that were meant to incite the squatters. Following that, a battalion was deployed in the area and squatters were persuaded to stay. The local councils dropped the matter because they could not adjudicate against the government decision that had ordered the ranchers to allow the squatters to get water from the ranches for their animals. The ranchers defied this decision.[112]

It was agreed during the ministerial fact-finding committee meetings between ranchers and squatters that the ranchers would desist from preventing cattle keepers or squatters from watering their animals at their water points. This was one of the major causes of the breakdown of security on the ranches. The squatters on the other hand, were required to dip/spray their animals against ticks and to install taps on valley tanks so that cattle would not leave to go into the valley tanks to drink water.

The Mobilization of the Squatters

It has been rightly argued that the squatters were highly organised in their uprisings. There were very few communal water points in the schemes. Watering was done in the afternoons. On any one afternoon, there would be several herds at one water point. Originally these were moments when the squatters discussed where to find more water and pastures. During the preparations for these uprising, their strategies were discussed there. This made co-ordination easy. The mobs of squatters attacked ranches, cut down perimeter fences, and forcibly grazed and watered their animals. Instead of watering animals in the afternoon, they started watering them in the mornings. They would take ranch owners by surprise by attacking ranches at dawn when they were unexpected. Reports indicated that the herdsmen had been aided, assisted and armed by their sons or relatives in the NRA, to attack and seize by force the ranches owned lawfully by large scale ranchers who had occupied them since the 1960s.

Where strong resistance was expected from ranchers that were known to possess guns, the squatters would split up into several groups. One group would divert the attention of the ranchers from water areas and would then disarm the ranchers, as the other groups watered their animals. The latter groups comprised mostly young girls and old men. All squatters affected by the water crisis participated irrespective of their ethnicity. The young and old participated. The young did the actual fighting. The old inspired the fighting spirits in their youths by recounting to them how much suffering they had endured under the ranch owners.

The fencing off of ranches by the ranchers forced the squatters to cut through the fences in search for water for their animals. The sheer number of squatters usually overwhelmed the security guards deployed by the ranchers. The squatters were said to have captured a total of ten guns. The ranchers could, in most cases, offer little resistance. They simply fled. At some point it was even alleged that the squatters had undergone some basic military training in the period preceding the uprisings. After a ranch was overrun, the squatter would then proceed to another. Because they expected the ranchers to reinforce, the squatters sealed off captured areas with road barricades. The squatters carried spears, pangas, clubs and sometimes guns.

The Degree of Violence

The magnitude of the violence varied on the ranches. It was highest at the Masaka ranches and almost non-existent on the Bunyoro ranches. On the schemes that experienced clashes, the degree of violence varied between individual ranches on the particular ranching schemes. The variation in the level of violence depended on the following:

a the level of government infrastructure development prior to the allocation of the ranches;

b the kind of relationships that had emerged between squatters and ranch owners;

c the availability of water in the areas surrounding the ranches;

d how developed water and pastures resources were on particular ranches.

Depending on how much the squatters believed that the ranch owners had accumulated wealth from their (the squatters') sweat was reflected in the nature and extent of damage inflicted on the properties of the ranch owners. The losses are not easily quantifiable in monetary terms. But the following complaints contained in a memorandum from the Marumba Ranching Co-operative on Ranch number 39 in Ankole gives a picture of the extent of damages. Writing to the Minister concerned, the Chairman of Marumba Co-operatives said squatter invasion between August 17, 1990 and January 1991 had led to:

> ... excessive ticks and tick-borne diseases, as new comers were neither dipping nor spraying; introduction of diseases hitherto unknown on the ranch, e.g. CBPP, Lumpy skin, eye diseases, udder diseases; scramble for water and pastures; clogging of valley tanks since cattle were walked into the water; cutting of the perimeter fences to allow their easy entry onto the ranch from all angles; vandalisation of infrastructure in order to force the ranch owner out; adulterating acaricides in dip tanks by adding excess water which led to loss of cattle due to ECF despite dipping thrice instead of twice a week; losing heifer calves; theft of bananas; in August 1993, workers were chased away by resistance councils and security personnel; for over a week, during which period the ranch headquarters were broken into and animal drugs, equipment, cash and other items were stolen.[113]

The squatters vented their anger in different ways. There had been a uniform pattern of destruction of property. Doors and windows had been destroyed, and in some instances movable property had been taken out of the houses and set ablaze. However, no human life was lost according to the findings of the delegation.[114] Squatters destroyed property on the ranches.

Take the following examples: Ranch No. 9, of Masaka Growers Co-operative Union, was attacked on 6 August, 1990, and the following property was destroyed. Huts were set on fire. Managers' houses, offices, and stores were broken into and all facilities including files of the last 22 years were set on fire. Drugs, acaricides, barbed wire and fences were destroyed. The dip tank's side roof was destroyed and arcaricides poured out. Soil and stones were all thrown into the dip. The number of stolen or killed animals could not be established due to security reasons.[115] In the same report it was said that on ranch number 14 all the buildings were destroyed including all the household items. Literature, drugs and acaricides were set on fire. Some parts of the water engine were dismantled and taken. Diesel fuel in the store was used to burn property. All the glass in the doors and windows was smashed. Animals were dying. Four calves had already died, and 10 others were evidently sick.[116]

On ranch No. 15, huts for herdsmen were set on fire. An engine was destroyed. In the manager's house, all furniture, records, drugs and bedding were set on fire. On this ranch, many squatters could be seen grazing and watering their animals immediately after the violence. On ranch No. 47, the main house was attacked, and bullets had smashed through the windows. A new engine and 10 rolls of barbed wire were missing. All household property had been destroyed. Over 20 goats had been slaughtered and eaten. A gang of raiders was still camping on this ranch and the remaining herdsmen had been warned to leave the ranch at once or they would be killed.[117]

The following ranches were also attacked in a similar manner: 11, 12 (which belonged to Haji Katongole); 13 of Gubaala Ranchers Ltd.; 17 and 18 of Lugobe; 19 of Mr. Y.N.N. Ssenkungu; 20 of Mrs. Kayemba (widowed); 21 of M/s Kugumikiriza Ranchers; 22 and 23 of Masaka Co-operative Union; 26 of Aniyaliamanyi Ranchers; 28 of Gerald Ssemogerere; 36 of Michael Mulindwa; 37 of Kabunduguza and Sons; 42 of Gregory Kalutebwa, 45 of Mr. Kasujja and 58 of M/s Kisambwa Ranchers Ltd.. 'What was amazing, in all these events, was that the resistance councils, the district administrators, the police and the military were told not to intervene so that the warring parties could sort out their differences even if property secured under government loans was being dismantled.'[118]

Between August and December 1990, there was an unprecedented increase in the number of cattle thefts from the troubled ranching schemes. On September 4, a man was arrested at the city abattoir with 12 heads of cattle, obviously stolen, bearing marks of INCAFEX Ranchers, in Masaka.[119]

Around mid-August 1990, the government was finalising arrangements for the possible return to Rwanda of Rwandese refugees.[120] We were not able to establish the exact connection between the government's planned repatriation of Rwandese refugees, their alleged involvement in military training and subsequent clashes in the ranching schemes, and the subsequent invasion of Rwanda in October 1990. However, the council member for Mawogola county, Masaka said in the NRC that squatters in Mawogola were receiving military training. The council member for Kyaka county in Kabarole district also alleged that in the Kyaka refugee settlement scheme, the number of refugees had gone down from 30,000 to 10,000, and that 20,000 had been recruited into the army. Knowledge of any recruitment into the army of Rwandese refugees was denied at all levels of the government, as was the alleged training in Mawogola of Rwandese. It was said that the people who were being trained were local defence forces (LDFS).[121]

2.9.5 The Repossession of Ranches in Government Ranching Schemes

When the National Resistance Council (NRC) sat to deliberate on the report of the Commission of Inquiry into the ranching schemes following the outbreak of violence, a number of policy issues were discussed. The commission's recommendation to repossess only those ranches, which had performed poorly, had originally been upheld by the government. At the time it that it was discovered that some ranches failed to honour the conditions given to them by government during their initial allocation. Of these ranches, 53 were to be repossessed. Only 3 of these ranches were in Ankole, 19 ranches in Masaka, 9 ranches in Singo, 2 ranches in Buruli while 20 ranches were in Bunyoro (Ranches Restructuring, 1997).

In Ankole where ranching schemes began, the government provided almost all the requisite ranching infrastructure including the following: perimeter fencing, valley dams, and dip tanks. Those who were allocated the ranches tried hard to fulfil the development conditions set up by government. After Ankole, the government established the Masaka ranches. Here the government provided up to 50 percent of the required infrastructure. By the time the ranches were established in Kabula (Mawogola and Lwemiyaga), Singo and Bunyoro, those who had been allocated ranches had to do everything themselves. Ranchers from Masaka and other ranching schemes said the government was being sectarian; that was why it was reducing the number of ranches in Nyabushozi to three. When the issue was put to the cabinet, some cabinet members sympathised with the point of view of the Masaka ranchers.

The government went back to the cabinet in the light of the innuendoes that it was targeting only ranches in Masaka and not those in Ankole and recommended that all ranches be reduced across the board. The matter was tabled to the National Executive Committee (NEC) which endorsed the government's decision to have the ranches reduced to three (for the good performers), two (for moderate performers) and one (for those who had performed poorly, but were still interested in ranching, and the majority of the squatters). After this decision had been made, people from Lwemiyaga and Mawogola again told the President that in their view, the government's previous decision to repossess only ranches that had performed poorly made more sense.

The government stood by its decision. The proposal in the Commission's report was revised because it had a problem regarding the criteria it used for determining non-performing ranches. The Commission of Inquiry used only one criterion, the number of stock. This criterion was irrespective of how the ranchers had performed with regard to other useful indices of performance. Building perimeter fences, setting up dips, developing water reserves, building paddocks, clearing the bush, were other indices of performance.[122] The argument that Ankole ranches had been more developed than those in Masaka and Mawogola would not have been true if the criteria of ranch development had been widened to include the other relevant parameters that show performance.

The same people again complained that the first decision had been better. A series of high-level government meetings were convened. The National Executive Committee (NEC) met on 14 July and 15 July 1990 to review the recommendation of the government commission of inquiry, and decided to uphold government's policy. A ministerial statement was issued to that effect. Between January and December 1989, tensions continued to build as alliances and counter-alliances were formed between ranchers and squatters on the one hand, and with their sympathisers in the policy making organs, government departments and especially the military, on the other hand.

The ranchers had a very strong lobby. Many argued that they had a right to compensation, even those who had not developed their ranches or those who had been politically allocated ranches that they had not developed. There were also those in political circles that felt that ranchers who had appropriately developed their ranches should not be affected by the restructuring. Unfortunately, the majority of the latter category was in Ankole. Some started arguing that the government was subdividing ranches outside Ankole to give land to Banyakole. The government then decided that all ranches be sub-divided. On 25 January, 1990, the government announced that it had repossessed 62 ranches in Ankole, Masaka, Singo, Buruli and Bunyoro.[123]

2.9.6 The Restructuring of Government Ranching Schemes

The Justification and Objectives of the Restructuring

The uprisings played into the hands of the government. It provided the government with 'an opportunity to recast this whole policy for the better and remove, once and for all, the friction between the ranchers and the squatters, as well as ensuring the introduction of uniform disease eradication methods for all the cattle in these areas'.[124] The NRM government publicly criticised previous governments for having created the ranching schemes without thoroughly understanding the consequences of the policies to all sections of the population and the economy. President Museveni himself had been a front-runner in the struggles to eliminate nomadism among his own Bahiima people. Starting in 1966, after completing high school, Museveni, together with several colleagues (most of them now dead), led a struggle to educate nomads about modern husbandry practices and to resist being displaced from land designated for the establishment of ranches (Museveni, 1997).

The squatter uprisings had prompted the government to convene a special session of the National Resistance Council (NRC) to generate a consensus over what was to become the government's policy on the government sponsored ranching schemes. In four days of closed sessions, the NRC deliberated on the issues of the government sponsored ranching schemes. It was on the fourth day, Friday 24 August, 1990, that the NRC resolved that the government should repossess and restructure all ranches, which had not been developed in accordance with the terms and conditions of the allocation. The NRC resolved that the registrar of titles, cancel all titles in respect of government sponsored ranching schemes. This would be done on the grounds of non-fulfilment of the terms and conditions of ranch allocation. Fresh titles would be issued on the recommendation of the Ranches Restructuring Board to ranchers whose ranches would be reduced and to squatters within the ranching scheme who would be allocated land.[125]

The Minister of Agriculture Animal Industries and Fisheries on 27 September, 1990 announced that President Museveni had constituted a nine-man member board to restructure the government sponsored ranching schemes in the Central, Southern and Western regions. The names of the board members, and of the technical committee appointed to assist it are appended. This Ranches Restructuring Board (RRB) was created under general notice No. 182 of 1990, by the President in conformity with the NRC resolution dated 24 August, 1990.[126] The Board was required to implement recommendations made by a special session of the NRC that discussed the issue of ranches between 22 and 24 August, 1990, and recommend to the government means of improving the ranches in order to ensure sustainable livestock production. The board would scale down the ranches to three, two and one square mile to resettle landless squatters and their livestock in those areas.[127] The terms of reference of the board were appended.

Ranch restructuring was seen as a prerequisite condition for the transformation of livestock production from being predominantly nomad-based (different management; extensive, low management input) to commercialised forms of livestock production managed semi-intensively by former landless cattle keepers living on smaller numbers of animals. This shift in emphasis was the direct result of the failure to develop the livestock industry on a model that was based on the ranching philosophy of high management input, high selectivity, leading high output. Ranch restructuring was especially considered as a lasting solution to the constraints that had led to the failure of the former government sponsored ranching schemes started during the mid-1970's until the 1980's, such as squatting by landless cattle keepers and nomadic herdsmen. The government believed that as soon as its policy on ranches came into force, the problem of squatters would be solved once and for all, as the Ranches Restructuring Board would enable landless squatters who had cows to gain access to land.[128]

The government saw cattle keeping nomads as the greatest obstacle to commitment to develop the livestock industry. The government had been nursing the idea to make nomadic cattle keeping a criminal offence.[129] A Minister of State for Agriculture, Animal Industries and Fisheries, in charge of Anti-Nomadism abolishing nomadic cattle keeping and Water Development was appointed in December 1994, and served until the May 1996 presidential election after which cabinet positions were restructured. Compulsory animal health measures were also to be introduced. Ranchers and settled former landless cattle keepers would be forced to dip their cattle, or be faced with prosecution.[130] The then Minister of Agriculture, Animal Industries and Fisheries indicated that after the ranch restructuring, animal safety would be ensured by making public the status of each herd in the different kraals. The government was going to enforce all the regulations on modern ranching and farming. It was going to check irregular movement of animals and their products in order to stop the spread of diseases that could wipe out the national herd. Squatters and ranchers would be required to fence their pasturelands, clear bushes and effectively occupy their areas in order not to attract other who wished to encroach on their property.[131]

The Beneficiaries of the Ranches Restructuring

Government specified the categories of beneficiaries who would be allocated land by the Ranches Restructuring Board. The beneficiaries were expected to be people who had no land at all, but had cattle and were living as squatters on government ranches and on private leasehold. Those who had land elsewhere and sold it recently, with intention of getting free land through the restructuring process were ordered to move away from the ranches that were being restructured.[132] The beneficiaries had to be Uganda citizens.[133] No bureaucrat or member of military was expected to benefit.

Despite the strong desire of the government to prevent non-squatters (or the non-landless) from benefiting, not sufficient monitoring mechanisms were put in place to enforce such a directive. The government decreed that the ranchers and the squatters remain at the locations where they had settled when the ranch restructuring exercise commenced, while a solution was being worked out to end their conflicts. Any kind of movement by squatters from one ranch to another was strictly forbidden. Local working committees were established for each of the ranching schemes, composed of two representatives of the Board and the representatives of the ranchers, squatters and local administration and resistance councils. These local working committees were expected to make a daily study of the existing problems and take the necessary steps to prevent problems getting out of hand. However, these local working committees could not adequately monitor the movement of cattle keepers with their herds especially in the dry season.

In order to determine the citizenship of the would-be beneficiaries, the Board designed a citizenship verification form for squatters, specifically to identify those who were and those who were not citizens of Uganda. The facts indicated in the registration forms were cross checked with local resistance committees in the ranching schemes, as well as with elders and chiefs. When people claimed to be citizens and chiefs and local council committees knew them not to be so, the board would interview them. Ranch owners or other cattle keepers who had acquired land titles and whose lands were being subdivided were never subjected to citizenship tests because it was assumed that they were appropriately screened using the usual methods before they had acquired the land titles.

And as far as the Board was concerned, the citizenship test applied only to squatters. From our findings, no foreigners seemed to benefit from the restructuring exercise. Our findings indicate that the majority of the local council committees were comprised of squatters because they simply happened to have been the majority in these areas. The Board unknowingly was confronted with squatter-dominated local council committees to decide the fate of fellow squatters. Even when there were non-squatters on the committee, the fact that this screening exercise occurred after the squatters' uprising indicates that not being a citizen was unlikely to be mentioned, especially if it involved some of the feared squatters.

Our findings also indicate that being a foreigner was not the major source of contention in the citizenship issue, but double citizenship. In Uganda, a number of ethnic Tutsi's of Rwandese origin claimed citizenship in Uganda as Ugandan Banyarwanda without necessarily giving up their other citizenship. Among those who benefited from the ranching schemes there were a number of this category of people who called themselves Banyarwanda. The Board had no mechanisms of verifying double citizenship.

2.9.7 Rationalisation of Private Ownership

Land Allocated to Former Ranch Owners in Ankole

Some ranchers retained much of the land they initially held before the restructuring exercise but there were a few ranch owners who completely lost all land. Of the 50 ranches in the Ankole ranching scheme, only one ranch owner (ranch number 22) retained all his land. 48 ranchers lost some of their land. Out of the total land area available for redistribution in the ranching schemes, the former squatters were left with 45 percent of the land in Ankole ranching scheme. 38 ranchers (76 percent) retained more than half of their former ranches, while only 16 percent could keep less than half their former ranches. Only 8 percent retained more than 80 percent of their ranches, as shown in table 4.

Table 4 *Amount of land retained by ranch owners*

Percentage of total ranch land Retained by former ranchers	*Number of former Ranch owners*	*Percentage of total*
0 %	1	2
1-19 %	2	4
20-39 %	5	10
40-59 %	19	38
60-79 %	19	38
80-99 %	3	6
100 %	1	2
Total	50	100

Source: Analysis of records of the Secondary data from the RRB

A detailed analysis of the amount of land retained by ranch owners compared to that retained by former squatters is shown in table 5.

Land Allocations to Squatters

In the criteria that was used by the RRB to sub-divide land in the ranching schemes, every individual head of a household registered as a genuine squatter was supposed to receive a minimum of 30 acres. To every head of cattle owned by such a squatter family, one acre was added, but did not exceed 270 acres. In other words, no squatter family would receive more than 300 acres. While the ceiling of 300 acres was never exceeded for squatter families, one squatter family ended up with an allocation of only 30 acres, the basic minimum earmarked at the beginning of the restructuring exercise. What this meant was that this household was registered as a squatter household, although they did not have any cattle! An analysis of the amount of land allocated to each squatter in Ankole RANCHING scheme is given in table 6.

From the criteria of land allocation used by RRB, each of the categories above corresponds to a specific number of animals registered for each squatter household at the time of the ranch restructuring. Therefore, it can be inferred that those who got only the minimum 30 acres did not have any cattle. Those who got between 31 and 59 acres had between one and 30 heads of cattle. Those squatters who were allocated between 60 and 89 acres had between 31 and 60 heads of cattle, and so on. The implication of the above is that, among the squatter families allocated land in Ankole, there were several of them that were actually cattle-poor pastoral households.

Land in the restructured ranches was allocated to the head of a family, who would hold the land in trust for his family members. The number of squatters settled on each ranch varied from ranch to ranch. The highest number settled was 36 squatter families on ranch number 29, while no squatter family was settled on ranch number 22.

Improvement of Land Allocated

The land that was allocated to each family was never demarcated or surveyed immediately because of the following reasons:

i The RRB did not have the required resources to carry out such work;

ii The RRB noted that some units could not be self-sustaining given the conditions in the area. The demarcation of land units for each individual family would leave some families without suitable sites for water let alone pastures.

iii The RRB recognised the fact that those allocated land were still poor and relatively backward, and it would therefore take time to adopt commercial farming. This necessitated the government to take the responsibility to pro-

TABLE 5 *Land Allocations in the Ankole Ranching Scheme*

Ranch No.	*Total area (ha)*	*Amount Retained by ranchers (ha)*	*Percentage Retained by ranchers (%)*	*Number of squatters settled on the ranches*	*Amount retained by squatters (ha)*	*Percentage retained by squatters %*	*Average land allocated per squatter (ha)*
1	1406	1035	73.6%	13	371	26.4%	28.5
2	1430	783	54.8%	25	647	45.2%	25.9
3	1570	801	51.0%	21	775	49.4%	36.9
4	1107	790	71.4%	13	317	28.6%	24.4
5	1213	767	63.2%	12	446	36.8%	37.2
6	1563	1036	66.3%	10	527	33.7%	52.7
7	1688	776	46.0%	18	912	54.0%	51.0
8	913	513	56.2%	13	400	43.8%	31.0
9	1173	770	65.6%	11	403	34.4%	36.6
10	1507	1016	67.4%	13	491	32.6%	38.0
11	1439	521	36.2%	30	918	63.8%	30.6
12	1331	1035	78.0%	9	296	22.2%	33.0
13	818	246	30.1%	21	572	69.9%	27.2
14	1222	786	64.3%	11	436	36.0%	39.6
15	1342	1036	77.2%	10	306	22.8%	30.6
16	1406	1147	81.6%	34	259	18.4%	7.6
17	1310	776	59.2%	14	534	41.0%	38.1
18	1615	777	48.1%	21	838	52.0%	39.9
19	1527	773	51.0%	21	754	49.4%	35.9
20	1473	777	53.0%	25	696	47.2%	27.8
21	1324	1036	78.2%	7	288	21.8%	41.1
22	1322	1322	100%	0	0	0%	0
23	1039	516	50.0%	27	523	50.3%	19.4
24	1428	778	54.5%	15	650	46.5%	43.3
25	1334	775	58.1%	10	559	41.9%	55.9
26	1619	0	0%	37	1619	100.0%	43.8
27	1196	518	43. %3	23	678	57.0%	29.5
28	1238	776	63.0%	22	462	37.3%	21.0
29	1882	271	14.4%	36	1611	85.6%	45.0
30	1472	747	50.7%	20	725	49.3%	36.2
31	1345	517	38.4%	21	828	61.6%	39.4
32	1447	259	18.0%	33	1189	82.2%	36.0
33	1701	1063	62.5%	13	638	37.5%	49.1

34	1218	787	64.6%	15	431	35.4%	28.7
35	1131	777	68.7%	12	354	31.3%	29.5
36	1426	777	54.5%	17	649	45.5%	38.2
37	1177	777	66.0%	16	400	34.0%	25.0
38	1675	785	47.0%	27	890	53.1%	33.0
39	1556	1054	67.7%	16	502	32.2%	31.3
40	1730	1033	59.7%	12	697	40.3%	58.1
41	1676	519	31.0%	37	1157	69.0%	31.3
42	1269	769	61.0%	15	500	39.4%	33.3
43	1143	1012	88.5%	(3)	131	11.5%	44.0
44	1526	776	51.0%	(12)	750	49.1%	62.5
45	1105	1035	94.0%	(2)	70	6.3.0%	35.0
46	1105	579	52.3%	(13)	586	53.3%	45.0
47	1376	791	57.5%	12	585	42.5%	49.0
48	1635	519	31.7%	22	1116	68.3%	50.7
49	802	516	64.3%	7	286	36.0%	41.0
50	1884	834	44.2%	25	1050	56.0%	42.0
	68,834	37,992	55.2%	872	30,821	45.0%	35.3

Source: Analysis of records of the Secondary data from the RRB

TABLE 6 *Land allocated to squatters*

Amount of land allocated to squatter households in Ankole	*Number of squatter Households and percentage of total*	*Approximate number of heads of cattle owned at the time of the census*[134]
<30	2 (0.2%)	Zero
30- 59	231 (26.5%)	1- 30 heads of cattle
60- 89	316 (36.2%)	31- 60 heads of cattle
90-119	152 (17.4%)	61- 90 heads of cattle
120-149	73 (8.4%)	91-120 heads of cattle
150-179	36 (4.1%)	121-150 heads of cattle
180-209	21 (2.4%)	151-180 heads of cattle
210-239	11 (1.3%)	181-210 heads of cattle
240-269	7 (0.8%)	211-240 heads of cattle
270-299	7 (0.8%)	241-270 heads of cattle
>300	16 (1.8%)	271-300 heads of cattle
Total	872 (100%)	

Source: Analysis of records of the Secondary data from the RRB

vide infrastructure such as valley tanks, dams, roads and so on. The argument was that groups rather than individuals would optimally utilise such infrastructure.

This demarcation started at the end of 1996. Everyone who had been allocated land was required to pay a premium and ground rent plus all other charges towards securing the titles and maintaining the basic social services. They were required, for the first five years, to pay premium per square mile of shs. 1,000,000 and ground rent per annum per square mile of shs. 100,000.00. Such payments were considered a way of making those who had been allocated land develop a strong feeling that the land was theirs. It was argued by the RRB that free things tend to get abused, fought for and tended not stimulate development. Payment of premium and ground rents would make those allocated the land more committed to developing and utilising it. It was also assumed that after paying premium and fees for the land allocated, those who had too many animals than the grazing capacity of the land would allow, would have the incentives to sell some of their animals, thereby reducing them to the desired number (RRB, 1997: 52).

538 squatter families (62.7 percent) allocated land on the restructured ranches were actually cattle-poor households who owned less than 60 heads of cattle at the time of restructuring. A total of 233 squatters, mainly cattle-poor households were off-loaded onto the Singo ranching scheme, were they were advised to settle down and practice mixed farming (Ranches Restructuring Board, 1997: 47). Although the majority of cattle-poor squatters were relocated to Singo ranching scheme, many remained in Ankole. These households would find it difficult to invest individually in the provisions of ranch infrastructure because of the huge capital expenditures involved. This is the reason why there was a minimum of investment in ranch development infrastructure in this initial post-restructuring period in Ankole. Even among the former ranch owners who were allocated land, many had not yet started investing in ranch infrastructure. On ranch number 9, the ranch headquarters located on that part of the former ranch that had been retained by the ranch owner, had dilapidated structures. There were not yet any indications of his intention to rehabilitate the ranch.

However, the most common activities through which cattle keepers were reducing unpalatable grass such as *Cymbopogon nardens* (*omutete grass*) and tree species such as the *Acacia hockii*, which had heavily encroached on pasturelands due to years of neglect, included the following:

i Cutting down the trees for use as building poles and fence poles. The cattle keepers allocated land in the restructured ranches are erecting new fence posts.
ii One land use activity that is increasingly on the rise on the restructured ranches is charcoal-burning and fire-wood selling. It is mainly the non-cattle keepers who carry out charcoal burning and collect fuel wood, which they sell on emerging trading centres along the Masaka-Mbarara highway. The cattle keepers who were allocated land, sell the right to cut trees for making charcoal at a given location to the charcoal burners. These are usually areas that have had a high encroachment of bush and trees. The types of trees, which are usually offered for charcoal-burning include the following tree species, Acacia, Combretum and Terminalia. Once wood for charcoal has been taken away, whatever has been left behind is sold as fire-wood. As charcoal burners cut down trees to make charcoal, the cattle keepers benefit two-fold. First, they are paid for the trees that have been cut down. Secondly, the cattle keepers do not have to spend any money hiring labour to cut the trees in order to plant more nutritive pastures. Some cattle keepers had hired labour to clear areas of trees and bushes, especially to uproot the *omutete grass*.
iii *Cymbopogon nardens* is mainly reduced by uprooting.

Politicisation of Ranch Restructuring

Former landless cattle keepers became landowners in the former Ankole ranching scheme overnight after their status became elevated by a political decision. This has created a situation of uneasiness because some of the former ranch owners retained small portions of their ranches. The ranchers were compensated, not for the land that was taken away, but for the developments they lost on the ranches. These developments included valley tanks, dip tanks, existing fences, spray races, buildings, bore holes and water pipes. Some of them have, however, continued to intimidate former squatters intimating that if the government changed hands, they would certainly loose the land they were allocated. For example, Bonifance Byanyima who lost up to 1291 acres (527 hectares) of land to ten squatters, retained 1036 hectares of land out of the original 1563 hectares on his ranch number 6. He said once: 'Museveni gave away part of the 5 square mile wide ranch which I started in 1959 to Rwandese squatters in July 1990. They beat me up, tried to take me to a military detach at Sanga, but I stood my ground. These Rwandese hold guns as walking sticks, because of Museveni's support.'[135] With such threats, very few cattle keeper would be willing to commit all their resources to improve portions of land they were allocated.

Emerging Forms of Landlessness

The rationalisation of private land rights in range land resources cannot be regarded as the only solution to resource management problems created by the failure of private land holding in the former ranching schemes. The restructuring and allocation of ranches to former landless cattle keepers has not led to the eradication of landlessness. There are some cattle keepers who remained on those parts of ranches which were not restructured, such as areas that were retained on ranches for the future construction of dams and valley tanks or service centres, or on government ranches such as Nshara dairy ranch in Ankole. Large pieces of land remained on almost all the ranches. The sizes of the land that remained were as follows: in Ankole, the Ministry of Agriculture gave up block 4 of the former Nshara dairy ranch for restructuring and retained block 3 which was meant to be developed as a model government dairy ranch. It is on these pieces of land that squatters who were not allocated land during the ranch restructuring have remained as a new breed of landless cattle keepers (squatters).

Some of the pastoralists who were not allocated land in Ankole moved to other ranching schemes where land subdivision and allocation had not started as early as it had in Ankole. They had fewer animals than the minimum number required to qualify for allocation. While the majority of these pastoralists actually moved on their accord, many of those who had applied to the Ankole ranch restructuring committee were referred to other ranches such as the Singo ranching scheme. The district authorities in Singo, following earlier Presidential directives on the subject of bona-fide squatters who would benefit from the allocation of land on the restructured ranches, refused to allow new entrants from Ankole and other areas to take up land on the Singo ranching scheme. This led to the proliferation of some new forms of landlessness. Those who lost out when Ankole ranches were subdivided could not be accommodated elsewhere in places like the Singo ranching schemes where they were rejected. In Buruli and Bunyoro, on the other hand, allocation was mainly based on either how long one had been squatting on the land, or how deeply rooted ones historical claim was to the land on which the ranches were established.

This means that there are a large number of landless cattle keepers whose rights to resources they previously had access to before ranches were restructured, have been greatly constrained by the recent subdivisions and land re-allocations. When the ranches were subdivided, the ranch owners were given an opportunity to decide which part of the ranch they would be interested in retaining. In the majority of cases, ranch owners chose portions of their former ranches that were more endowed with resources than the remaining parts. These were usually areas that had watering points, dipping facilities and better pastures. Squatters were allocated areas where they found themselves in a position of always having to depend on the resources retained in the part of the ranch that was retained by the ranch owners. Developing new facilities like water points is not easy. The costs involved are exorbitant. Where settlers have not pulled together resources to construct water points on ranches where water is seriously lacking, settlers have had to rely on renting either water or pastures from either the ranch owners or elsewhere from wherever these could be obtained. This is a form of landlessness that is defined by resource dispossession as well as prohibition of access to previously accessible resources.

Land Fragmentation

Ranch restructuring has greatly contributed to the fragmentation of land whose soil productivity (a function of soil structure, texture, and rainfall patterns) is considered marginal as has already been hinted upon in preceding discussions. This notwithstanding, it has not helped to stem conditions that have led to some of the problems that have been associated with cattle keeping in the ranching schemes before the restructuring of the ranches. From 50 former ranch owners, land ownership in the former Ankole ranching scheme has been transformed to 918 private landowners (including 872 former squatters and 49 former ranch owners).

2.10 Livestock Sector Development: Current Prospects and Future Constraints

2.10.1 The Development and Management of Water Facilities

Water Development by the Livestock Services Project

The approach to water development by the Livestock Services Project (LSP) has involved the construction of new valley tanks and dams and the rehabilitation of existing ones using the District Veterinary Department. Here water development was supervised directly by the Ministry of Agriculture, Animal Industry and Fisheries, and was largely considered to have failed to produce the desired result because of the minimal involvement of local pastoral communities.

During the 1997/98 financial year, the Ministry of Agriculture, Animal Industry and Fisheries said it had constructed 19 large, strategic water reservoirs in six drought-prone districts of the cattle corridor. There was one reservoir in Ntugamo. There were eight in Mbarara, two in Ssembabule, two in Nakasongola, three in Mubende and three in Kiboga. While considering the Ministry's budget and policy statement, a number of complaints had been made to the Sessional Committee of the Ministry. For example, the Member of Parliament for Lwemiyaga in Ssembabule explained that:

> The Rwamakara dam claimed to have been constructed by the ministry during the financial year 1997/8, was an old colonial dam built in 1948. The farmers had wanted the dam rehabilitated by removing dirt, silt and dirty water, at the time mixed with cow dung and suspected to be causing disease to their livestock. The Member of Parliament lamented that the ministry's contractors had just cut into the dam wall without removing any silt, cow dung or rotten plant material in the empty dam reservoir. They had laid pipes, built water troughs and increased the old dam wall by two meters, fenced the dam, and claimed their 99 million shillings and walked away.[136]

Other Members of Parliament made the same complaint. Old dams were being drained without de-silting them. The Nyabushozi Member of Parliament said that on Kishangara dam, Uganda Shilling 117 million was spent. Mubende Members of Parliament said that 178 million was spent on Dyangoma dam. All three dams did not have water, and yet nearly 400 million had been spent on them. At Nyakahita in Nyabushozi, 149 million had been spent on the site, and there was nothing that looked like a dam anywhere around the area. In the rest of the country where cattle rearing was a major economic activity, there was no coverage by this project.

The technocrats in the Ministry insisted that it would require an engineering audit to prove the allegations that there was no value for the money sunk into the water development under the Livestock Services Project. The following were cited as having caused the failure of the project to realise its original objectives: the Ministry did not provide adequate supervision of the project. There was limited input from the beneficiaries and other stakeholders; the project was poorly designed. The donors largely dictated the management of the project, how the borrowed funds from donors would be spent, and the actual supervision of the project. There was a lack of technical and political direction and capacity on the part of the ministry.[137]

Participatory Development of Water Facilities

The second approach was used in the Ankole Ranching scheme, where the Integrated Pastoralists Development Programme (IPDP), supported by the German Technical Co-operation (GTZ) mobilised the cattle keepers to participate in water development and management directly. This approach has largely been a success.

One of the major problems of livestock production is the availability of adequate water supplies. Many cattle keepers often express this problem. Once a group of cattle keepers identifies a need and expresses interest in overcoming the problem, the IPDP field staff convene a community meeting to determine who the potential beneficiaries will be. The number of users as well as the number of animals to be served by the water facility will be taken into consideration.

The community, of usually between 40 and 50 families that share a common water facility, is asked to constitute a task force to identify sites where to put the water facility. A feasibility study is carried out by the IPDP to determine how much water is needed and what kind of facility would provide the required amount of water (tank sizing). The costs are then calculated. The size of the water facility is determined by the people's willingness to contribute local resources for its construction and the population of livestock as well as human beings to be served by the facility. Before any work can begin, the community formally writes an application to the IPDP for assistance. The local community signs a contract together with the project management that specifies the particular obligations and responsibilities of each of the parties involved. The contract is signed by the Project Technical Adviser, the Sub-county Council (LC3) chairman, the Sub-county chief, and the representative of the cattle keepers and the Project Manager.

For much of the time, the project provides non-local material (cement, barbed wire, hand-pumps, pipes etc.) and other inputs required to make water available such as technical advise for trough construction. The local water users provide all the locally available materials such as bricks, sand, physical labour or stone aggregates. When money is required, to pay either for fuel or to motivate the drivers of the earth-moving equipment, for example, the amount to be paid usually depends on the number of cattle a family has. The cattle keepers' pay is fifty Uganda shilling (shs. 50) for each animal owned.

As the water facility is being worked upon, the task force is re-constituted into a Water-Users' Committee (WUC). The WUC ensures that all targeted users give their contributions. They ensure that once water is available, its use is sustainable. The WUC nominates a person who is trained to maintain the water structure. All the users are educated in the best way to use the facility. After the facility is ready for use, the community agrees upon bye-laws in consultation with the IPDP. There are six people on the WUC, including a Chairman, Secretary, Treasurer, Caretaker and two committee members. Although two of the members of this committee have to be women, it has often not been possible to find women who were interested in the areas where these committees set up. It is only on Ranch No. 7/11 where such bye-laws were in an advanced stage of development, and where they have set up a WUC.

2.10.2 The Increase in Cultivation on the Restructured Ranches

There is widespread cultivation taking place on land allocated in the restructured ranches, with bananas (*musa spp.*) being the most common. The following crops were also grown: sweet potatoes (*ipomea batatas*), maize (*zea mays*), cassava (*manihota esculanta*) and beans (*phaseolus vulgaris*). For a casual observation, the reason why cultivation was being undertaken is the need to supplement the immediate households' subsistence requirements. Many different households of settlers in the former ranching schemes were adopting the growing of crops for different reasons, including the following:
i The cattle keepers are able to sell *matooke* to buy other family requirements. To do so in the past, they had to sell livestock.
ii Everybody was planting *matooke*. It used to be difficult at the beginning because suckers had to be bought from very far. After the bananas have grown, the cattle keepers do not need to spend a lot of time looking after them. They do not require a lot of labour to tend. While food is then assured, it is then possible to release labour for other more demanding activities like looking after livestock.
iii Growing of a variety of crops increases food security for the household.

However, to understand the significance that this increased cultivation has for livestock rearing, the cultivation of crops has to be linked to other agricultural activities the cattle keepers are involved in. They are practise a land use system where they grow woody perennial trees on the same piece of land with food crops (especially bananas), and at the same time use the same piece of land for livestock rearing. The tree species that had been planted included the following:
a Cashew nut tree (*Azadirachta indica*); shea butter trees (*Butyrospermum porkii*); edible figs (*Ficus gnaphalocarpa*); mangoes (*Mangifera indica*); avocados, etc. These trees grow tall with open canopies that allows grass to grow right to their bases. Because they are evergreen, they prevent the total drying of grass under them, and also provide shade for the animals whose manure improves the fertility of the soil for better pastures.
b The cattle keepers have also planted deciduous tree species, which they use to divide their ranches into paddocks. Such species include: Acacia *nilotica*

(tannin); Acacia *senegal* (gum arabica) and Acacia *seyal* (gum arabic). A thorny shrub called Key apple (*Dovyalis cofra*) has also been planted. The combination of the deciduous trees and the thorny shrubs provides a permanent live fence.

c Certain species of eucalyptus trees have been planted to provide building materials (poles for construction of houses and fencing posts) and firewood.

d Forage trees and shrubs, which provide browsing especially during the dry season when the grass has dried, have also been planted. These mainly include edible figs (*Ficus gnaphalocarpa*).

e Cattle keepers have also planted tree legumes, which increase soil fertility because of their nitrogen fixing ability. These tree species include *Sesbania sesban* and *Sesbania grandiflora.*

f Cattle keepers have also planted fruit trees around their homesteads such as mangoes and avocados.

g Banana cultivation has provided a very important supplement of animal fodder as well as supplementing the household income for purchasing other household needs.

While it may appear that the food crops are grown for direct family consumption, they also serve another purpose, which enhances their ability to practice cattle rearing. The cattle keepers plant crops such as maize, beans, and groundnuts. Apart from their nitrogen fixing abilities, which supplement the effects of the tree legumes they also serve as standing fodder. Once the crops have been harvested, the crop stems are never cut down immediately. They remain in the fields and provide standing fodder used for feeding livestock. As livestock browses on these standing crop residues, they replenish soil nutrients with their urine and excreta. Once a mature banana has been harvested, the cattle keepers cut the stems into smaller pieces, which are scattered in the fields to provide mulch for other crops, and also as feed supplements for livestock.

The above is an indication that cattle keepers have achieved varying degrees of integration of trees, crops and livestock on the same pieces of land, which serve both ecological and economic functions beneficial to livestock production in general. The increase in cultivation is not so much an indication of a changing practice to settled farming or mixed farming, but an attempt to sustain cattle rearing further. This integration increases the productivity of the land, which helps to sustain it to a high level for a relatively longer period of time without degrading the total environment.

Integrating trees in livestock production is an important feed supplement for livestock. Tropical grasses have adequate protein content for only 4-6 months of the year (during the rainy season). As they mature, especially towards the dry season, they become fibrous and less digestible. During this period, browse from trees and shrubs provide the required protein for the livestock. It has been argued (in Ranches Restructuring Board, 1997: 60), that grass drops in protein to less than 6 percent, which affects the ruminant's microbes so that they cannot produce adequate protein to maintain their growth and reproduction rate. This affects the animal's food intake, which drops and ensures loss of conditions and decline in milk production.

2.10.3 Current Manifestation of Pastoral Nomads

At the time when the policy for restructuring government ranching schemes was being discussed, it was the strong belief of the government that pastoral nomads were the greatest obstacle to the development of the livestock industry. The general point of view was that soon after the end of the ranch restructuring exercise, a law banning the nomadic life style would be introduced in parliament making it an offence for cattle keepers to continue roaming from one place to another. It was also anticipated that compulsory animal health measures would also be introduced once former landless cattle keepers were allocated land. All beneficiaries of allocation of land on the restructured ranches in the former ranching schemes would be forced to dip their cattle, lest face prosecution.[138]

Similarly, the then Minister of Agriculture, Animal Industries and Fisheries indicated that after the ranches were restructured, animal safety would be ensured by making public the status of each herd in the different kraals. The government was going to enforce all the regulations on modern ranching and farming, to check irregular movement of animals and their products, and to stop the spread of diseases that could wipe out the national herd. Squatters and ranchers would be required to fence their pasturelands, clear bushes and effectively occupy their areas in order not to attract other encroachers.[139]

Cattle keepers have constructed semi-permanent houses and a few permanent houses, but they still have to move their livestock, mainly in search of water during the dry season, but also sometimes of pastures as well. The south-western parts of Uganda's cattle corridor experienced a drought that began in November 1988, reaching an alarming level in April of 1999, and continuing to August 1999. Many cattle keepers were forced to move in search for water for their animals during this period of prolonged drought.

Up to 200 families from Nyabushozi and Kazo county in Mbarara district moved their herds to Lake Mburo National Park, while cattle keepers in

Bukanga, Rujumbura and other counties in southern Uganda moved their herds to northern Tanzania's Kagera crescent. At the beginning of September, a few families returned to their homes after a short spell of rains. The rest were given a deadline ending at the beginning of October 1999 to vacate the Park. Because the rains had not been sufficient, this deadline was extended to December 1999. Many of those who moved to Lake Mburo National Park had built permanent homes and settlements where they return after the drought ended.

In Nyabushozi, many cattle keepers who had the means provided water for their animals in water tankers water tanks. In one such instance, a cattle keeper put a very large polythene sheet at the bottom of the valley tanks before filling it with water to prevent the water from percolating into the soils. There were some ranches that had no water points. During the dry season they moved their animals to places where they could rent the use of a water source. They did not move away completely as parts of their families remained behind. Those who moved with the entire families returned to the same place after the dry season ended. These cattle keepers practised a system called transhumance. It is quite misleading to consider the movement of these cattle keepers as a way of life of traditional cattle keepers. It has increasingly become a survival strategy of all categories of cattle keepers who lack critical pastoral resources during periods of critical scarcity, most especially the dry season.

Even after land was allocated, the cattle keepers continued to treck for resources on a seasonal basis. This is partly because the movement of cattle keepers with their livestock, otherwise referred to as nomadism, is not primarily caused by a lack of private ownership of land. You do not need to own land as private property to be able to have access to requisite pastoral resources. Owning land privately does not automatically mean one has all the required resources for livestock to last throughout the dry season. The movement of cattle keepers and their livestock can only be minimised if requisite resources such as water and pastures are available in sufficient amounts throughout the dry season.

2.10.4 The Development of Livestock Production: What is the Way Forward

A strong ranchers' lobby argued, during the debates on whether or not all the government ranching schemes should be repossessed and restructured, that the political and economic problems from the mid-1970's to the 1980's had had adverse effects on the overall economy and the national life, and had in turn affected the commercial livestock ranching sector (Republic of Uganda, 1988). It was argued that following the end of the civil war, had they been given another opportunity, the ranchers would have been able to rehabilitate and develop their ranches in accordance with the conditions specified in the lease agreements.

It was presumed in this argument that the principles upon which the ranches were established in the 1960's were still valid. This argument ignored changes in human and livestock population and a historical fact that when the ranches were set up, traditional cattle keepers who had occupied these areas had been displaced. The arguments advanced to counter these claims by the ranchers lobby, point to a problem underlying the very concept of commercial livestock ranching. President Museveni, reacting to the 'war-affected-the-ranches' claim during the debate on the ranches in August 1990 argued:

> ... there was no fighting in Nyabushozi. There was no fighting in Mawogola, there was no fighting in Kabula, even in Buruli, there was no fighting in Buruli proper. The fighting was on in Singo. And fighting of course in Aswa, in Acholi ... I have been asking a question to the ranchers, you say you lost your cattle because of the war, but how about the squatters, they have also been living here? Why did they keep their cattle and you lost yours, squatters and people who had no land, who are moving, who are harassed, who are what, they still keep their cattle, and while you who has got every facility you lose yours? Now the ranchers say, because we were having delicate breeds, meaning the exotic – okay what does that mean? That may mean that ... may be these animals were introduced prematurely![140]

The above implies that certain wrong assumptions were made about commercial livestock ranching as a solution to the problems of livestock production, and thirty years later these assumptions are still fundamentally flawed. This suggests that for a long time part of the problem of livestock development had been the (pre-mature) introduction of certain 'modern' methods that were not appropriate for the complexity of systems into which they are introduced. This complexity may arise from internal processes of change within these systems, but in this particular situation they were mainly the result of an external situation of uncertainty created by political insecurity. Not surprisingly, the traditional cattle keeping sector was better able to cope with these uncertainties.

It is true that traditional pastoralism has been changing significantly, and as has been advocated for by government, certain 'traditional ways' need either to be improved or transformed (Republic of Uganda, 1996). But at what point should these changes be introduced, so that in future, they are not judged as having been introduced pre-maturely? Increasingly, this points not only to the problem of conceptualisation, but also the problem of strategy. The reason why the ranchers lost all their animals while the squatters did not has been argued about extensively. It was not only the way in which the ranches were introduced – for that created a lot of friction between the squatters and the ranchers – but also the way in which strategy for animal development was adopted. Commercial ranching was based on large-scale cattle keepers as opposed to smallholder cattle keepers.

For livestock development projects to succeed they must not only avoid creating social disharmony, but they must also be socially acceptable for the different stakeholders, especially the cattle keepers in the traditional sector. They constitute a strong social power base because they are numerically in the majority. This means that livestock development projects should target them directly. To make this point Museveni related his own experience:

> As you know I have been hunted by two governments. I was hunted by the government of Amin, it was hunting me, they came to arrest my father and so on and so forth. I was also hunted by the government of Obote. I was a principal target. But I may be able to inform you that none of those governments has ever eaten my cows. Why? I am asking the ranchers to tell me. In fact, on two occasions, they came actually to eat my cows. And who hid the cows? My neighbours. But the ranchers, the problem is that as soon as there is a slight disturbance, a slight breakdown of law and order, because they have got enemies around them, and because the cattle are also delicate, all the cows are eaten by the time there is a war. So at the end of every slight disturbance we must sit down and say we must re-stock. We must restock, so you go to the Bank, you get a new loan, you restock. When you are beginning to restock, a slight disturbance, those are lost.[141]

It is not surprising that the NRM government has focussed livestock development policies on the small holder cattle keepers who are pastoral nomads. That was the logic of ranch restructuring – to make land available for resettling landless cattle keepers. However, the wartime period presented a Janus-faced experience for livestock production. Does the ability of squatters to safeguard their livestock better than the owners of ranches during the civil wars, suggest that traditional pastoralist production systems allow the cattle keepers to take advantage of survival opportunities wherever and whenever they arise? During wartime, they can protect the herds of those who are in danger by concealing the identity of their owners. When the security situation worsens, they move their livestock away from danger.

During the dry season, this movement, otherwise referred to as nomadism, is sometimes intended to enable herds to find more nutritive grasses in wetter areas. This movement away from less productive range-land areas during the dry season to wetter areas also allows the cattle keepers to prevent the degradation of the range lands. Sometimes on a daily basis, the cattle keepers are able, through a system of mobile grazing, to walk their animals over large areas of unproductive range lands and to retire to watering points at the end of the day. Faced with physical, social, economic and political uncertainties, that translate into production constraints to cattle keeping, it is unlikely to transform cattle keepers in a desirable manner, especially when it involves an end to continuous movement of cattle keepers and their herds, for whatever reasons this is to be undertaken. The way forward lies in strategies that are designed to tackle these constraints.

Notes

1 The argument that state intervention affects the functioning of traditional systems of resource management, and fails to replace them with new institutions, and yet the old ones are rendered irrelevant has been advanced, among others by (i) Lane, Charles, 'Past Practices, Present Problems, Future Possibilities: Indigenous Natural Resource management in Pastoral Areas of Tanzania,' in Marcussen (1993); (ii) Vedeld, Trond, 'The State and Commons in the Sahel: Observations on the Niger River Delta in Mali,' in Marcussen (1993); (iii) Bromley, Daniel, 'Resource and Economic Development: An Institutionalist Perspective,' *Journal of Economic Issues*, Vol. 3, September 1985, pp. 777-796.

2 See *Parliamentary Hansards*, issue No. 14, 28 June 1990 to 23 August 1990, pp. 388-389.

3 See 'Back Ranches Plan, Museveni calls,' *New Vision,* 17 November 1990, pp. 12.

4 In his recent biography, 'Sowing the Mustard Seed,' President Museveni says that as early as the 1960's when he was still a high school student, he tried to educate the Bahiima cattle keepers on modern ways. President Museveni also says he tried to organize them to resist the setting up of the ranching schemes (Museveni, 1997).

5 See *Parliamentary Hansards*, issue No. 14, 28 June 1990 to 23 August 1990, pp. 389.

6 Both secondary and primary sources of data were used during this study. Background information on the livestock production sector and livestock production policies were obtained from the Ministry of Agriculture, Animal Industry and Fisheries policy documents. Reports on specific livestock projects referred to this study have been referenced. Data relating to the development of livestock production in the past was obtained from the National Archives as well as MAAIF, Entebbe. Issues of policy relating to the livestock sector were obtained from Annual reports of the Ministry. To obtain primary data, I interviewed some officials of the Ministry of Agriculture, Animal

Industry and Fisheries. The District Veterinary Officers, Mbarara district, Nakasongola District and Masindi district and officials who were responsible for the ranches restructuring exercise in these districts were interviewed. I also interviewed field staff of GTZ/IPDP, Mbarara. During rapid rural appraisals, cattle keepers were interviewed in Bunyoro ranching schemes, Buruli ranching scheme and Ankole ranching scheme.

7 In File A44/19, UPSMP No. 159/1908: Exportation of Cows to EAP, National Archives, Entebbe.

8 UP Annual Report on the Veterinary Department for the year ended 31 December, 1951, Entebbe, Government Printers, 1952, National Archives, Entebbe.

9 IBRD, 1961. *The Economic Development of Uganda*. Entebbe: Government Printers.

10 See Colonial Office, 1955. *Report of the East African Royal Commission Report of 1953*. London, Her Majesty's Stationery Office.

11 Bruce, J.W. and C. Tanner, 'Structural Adjustment, Land Concentration and Common Property: The Case of Guinea Bissau,' in *Marcussen* (1993), whose views are shared by *Ferder and Noronha* (1987) and *Boserup* (1981), observed that strengthened market forces tend to cause indigenous land tenure to evolve (as if automatically) towards stronger individual rights and weaker community rights in land, and that market forces are strengthened by many factors, including commercialization of production, rise in trade in previously subsistence consumption products, etc.

12 Though Ault, E.D. and L.G. Rutman, 'The Development of Individual Rights to Property in Tribal Africa,' *The Journal of Law and Economics*, Vol. 22 (1), October 1979, were able to distinguish between communal and private rights, in apparent objection of the uncritical description of property regimes in traditional societies, their arguments are some-what over-stretched. They, for example, point out that, 'because an individual only has a communal right to a specific resource prior to its use, communal systems create incentives for individuals to convert resources from a form over which communal rights are exercised into a form over which private property rights may be exercised' (pp. 166). The argument is presented as if there is an overwhelming drive for individuals to privatize resources, and as if communal property rights in society are a historical accident. This may not necessarily be correct.

13 UP Annual Report on the Veterinary Department for the year ended 31 December, 1938, Entebbe, Government Printers, 1939, National Archives, Entebbe.

14 Address by the Vice President, and Minister of AIGF, Hon. J.K. Babiiha, M.P. at the opening of the RAC on 15 January, 1970, at 10,00 a.m., in File C.120, opened July 1969, Ranching Advisory Committee (RAC), MAAIF, Entebbe.

15 Progress Control Report No. 1, March 1967, in File No. C.108A Ankole/Masaka Ranching Scheme (Progress Reports), MAAIF, Entebbe.

16 See minute 2/66 on the Background to the Ankole/Masaka Ranching Scheme, in communication referenced 373/CHE E/5 of 25 July, 1966 from G.D. Sacker, Subject: Ranching Selection Board: Minutes of 3rd meeting held at the MAIGF on 15 July, 1966, in File No. C/A3 RSB, MAAIF, Entebbe.

17 In Ref. C.MAI.V/8/2/A/12 of 25 June, 1966, from E.B. Galukande PS, MAIGF, to Secretary for Planning, Entebbe, subject: Ankole/Masaka Ranching Scheme, the PS informed the Secretary for Planning that 12 ranches would be ready for occupation in 2-4 months, in File No. C/A3 RSB, MAAIF, Entebbe.

18 Minutes of 2nd meeting of RAC of 25 March, 1970, in File C.120, opened July 1969, Ranching Advisory Committee (RAC), MAAIF, Entebbe.

19 'A short review of Ranch Development and Cattle Density in Buganda Region,' by Regional Veterinary Officer, Buganda, Mr. A. Iga, presented to the meeting of RAC of 10 September, 1971, in File No. C.120 (a), RSB, MAAIF Archives, Entebbe.

20 G.D. Sacker, 'The Ankole-Masaka Ranching Development Project,' in *Sacker and Trail* (1968).

21 G.D. Sacker, Senior Livestock Improvement Officer, and Secretary Ranches Selection Board, ref. 87/2/Com.5 to Mr. P.E. Esabu, Chairman Ranches Selection Board: Minutes of 7th meeting of the Ranching Policy Advisory Board of 19th February, 1965, Min. 3/65: Allocation of Ranches, in File No. C/A3 RSB, MAAIF, Entebbe.

22 Ref. 135/6/Com.5 from G.D. Sacker of 23 March, 1956 to Members of RSB. Subject: RSB, in File No. C/A3 RSB.

23 Ref. Com.5 of 29 March, 1965, from G.D. Sacker to James Kangaho, M.P., Mr. J.B.K. Kahigiriza, Enganzi Ankole Kingdom, Hon. Grace Ibingira, and Hon. Basil Bataringaya, In File No. C/A3 RSB.

24 This was an East African Posts and Telegraph Administration Telegram by Mr. Kangaho to Secretary, Ankole Ranching Scheme, dated 25 March, 1965, in File No. C/A3 RSB.

25 C.MAI.M.17/12/2 from W.W. Rwetsiba PS, MAIGF of 23 March, 1969, to G.D. Sacker, RE: RSB, in File No. C/A3 RSB.

26 Ref. Com. 5, Secretary RSB Sub-Committee, to F.E. Esabu, Chairman RSB, of 29 March, 1965, subject: Report on the meeting of the Interviewing sub-committee, in File No. C/A3 RSB.

27 These guidelines are contained in minute 3/66 titled 'New Conditions which the RSB was to follow in Selecting Ranchers,' in communication referenced 373/CHE E/5 of 25 July, 1966 from G.D. Sacker, Subject: Ranching Selection Board: Minutes of 3rd meeting held at the MAIGF on 15 July, 1966, in File No. C/A3 RSB, MAAIF, Entebbe.

28 Min 12/66: Future applications for Ranches, in Minutes of 4th meeting of RSB, in File No. C/A3 RSB, MAAIF, Entebbe.

29 Ref. E/716 of 24 August, 1967 from K.K. Karegyesa M.P. to Secretary, RSB. Subject: Applying for a Ranch, in File No. C/A3 RSB.

30 Ref. C.MAI.V/8/2/A/73 of 20 September, 1967 from A.B. Adimola PS MAIGF to Senior Livestock Improvement Officer, U.F.S. Commissioner of Veterinary Services and Animal Industry: Ranches Selection Board, in File No. C/A3 RSB.

31 *ibid.*

32 See Doornbos, Martin and Lofchie, Michael, 'Ranching and Scheming: A Case Study of the Ankole Ranching Scheme,' in Lofchie, M. (ed.). 1971. *The State of the Na-*

tions: Constraints on Development in Independent Africa. Berkeley: University of California Press, (pps. 165-187). I am grateful to Prof. Martin Doornbos for having agreed to e-mail me a printout of this chapter, having failed to find an original one. I am also grateful to Prof. Doornbos, for having allowed me to access his archives at ISS, at the Hague, which contained a private collection of records about this particular subject.

33 *ibid.*, pp. 153.

34 Contained in 'Notes on a meeting held on 3 May, 1965, Parliamentary Buildings to consider Loan applications to USAID by the Government of Uganda (Confidential), attended by Hon. Dr. A.M. Obote, Prime Minister; Hon. J.K. Babiiha, Minister of Animal Industry, game and Fisheries, Hon. S.N. Odaka, Minister of state for Foreign Affairs, Mr. J. Kakonge, Director of Planning; Z.H.K. Bigirwenkya, Ps Ministry of Foreign Affairs and E.B. Galukande, PS MAIGF,' in File No. C120 opened 5 December, 1963, Ranching Policy Advisory Board (Ankole), MAAIF, Entebbe.

35 G.D. Sacker, Senior Livestock Improvement Officer, and Secretary Ranches Selection Board, ref. 87/2/Com.5 to Mr. P.E. Esabu, Chairman Ranches Selection Board: Minutes of 7th meeting of the Ranching Policy Advisory Board of 19th February, 1965, Min. 3/65: Allocation of Ranches, in File No. C/A3 RSB, MAAIF, Entebbe.

36 B/11 Esabu to G.D. Sacker: Minutes of 2nd meeting of the RSB held on 1 April, 1965, dated 13 April, 1965.

37 For the Co-operative joint venture, see B/11 Esabu to G.D. Sacker: Minutes of 2nd meeting of the RSB held on 1 April, 1965, dated 13 April, 1965.

38 For composition of the RSB, see communication from PS MAIGF, W.W. Rwetsiba to G.D. Sacker, esq. Senior Livestock Improvement officer, Mbarara, ref. C.MAI.M.17/12/2, in File No. C/A3 RSB.

39 Ref. C.MAI.V/8/2/A/81 of 5 October, 1967 from PS MAIGF, A.B. Adimola to Commissioner of Veterinary Services and Animal Industry, in File C/A3, RSB.

40 Ref. C/A3 of 30 August, 1967 from G.D. Sacker to PS MAIGF, subject Ranching Selection Board, in File No. C/A3 RSB.

41 Ref. C.MAI.V/8/2A of 1 July, 1969 from PS MAIGF to officer in charge, Ranching schemes, Mbarara, subject: Occupation of Ranches, in File No. C/A3 RSB. See appendix for composition of RAC.

42 Ref. C.MAI.V/8/2A of 10 July, 1969 from J.W.L. Akol, PS, MAIGF to Officer in charge RS offices, in File No. C/A3 RSB.

43 These terms of reference have been stated in communication of 23 July, 1969 ref. C.MAI.V/8/2A from J.W.L. Akol, PS, Ministry of Animal Industry, Game and Fisheries, to Commissioner of Veterinary Services and Animal Industry, RE: Ankole/Masaka Ranching Scheme: Ranching Advisory Committee, in File C/A3, RSB, MAAIF, Entebbe.

44 Ref. C.MAI.V/8/2/A/74 from A.B. Adimola, PS MAIGF to Secretary, Ranches Restructuring Board, subject: RSB in File No. C/A3 RSB.

45 Ref. 89/3/com.5 from G.D. Sacker Senior LIO, of 19/2/1965, in File No. C/A3 RSB.

46 See Mbarara 166. C.A/3 of 25 September, 1968 from G.D. Sacker, Secretary RSB to Y. Walusimbi and Sons, P.O. Box 14041, Mengo, in File No. C/A3 RSB.

47 See Mbarara 166. C.A/3 of 25 September, 1968 from G.D. Sacker, Secretary RSB, to Mr. B. Karuragire and Company of P.O. Box 105, Mbarara.

48 Ref. No Vet. 1/1 of 1 September, 1967, from Kihimba, Ankole Kingdom Government to Secretary, RSB, subject: Ankole-Masaka Ranching Scheme – Selection of Ranchers in Ankole, in File No. C/A3 RSB.

49 Minute 13/68, AOB, in Minutes of 9th meeting of the RSB held at MAIGF, Kampala, 1 October, 1968, in File No. C/A3 RSB.

50 See Mbarara 166. C.A/3 of 5 August, 1968 from G.D. Sacker, Secretary RSB to Mr. Z. Rwanchwende, P.O. Box 298, Masaka. Subject: Ankole-Masaka Ranching Scheme, in File No. C/A3 RSB.

51 Ref. Mbarara 166. Conf./P/Reports/. of June 24, 1969 from H.J.S. Marples, Secretary, RSB to PS, MAIGF, subject: Occupation of ranches, in File No. C/A3 RSB.

52 Ref. Mbarara 166. Conf./P/Reports/. of June 24, 1969 from H.J.S. Marples, Secretary, RSB to PS, MAIGF, subject: Occupation of ranches, in File No. C/A3 RSB.

53 Ref. Mbarara 166. Conf./P/Reports/. of June 24, 1969 from H.J.S. Marples, Secretary, RSB to PS, MAIGF, subject: Occupation of ranches, in File No. C/A3 RSB.

54 Ref. C.MAI.V/8/2A of 1 July, 1969 from PS MAIGF to officer in charge, Ranching schemes, Mbarara, subject: Occupation of Ranches, in File No. C/A3 RSB.

55 See letter dated 17 July, 1973 from E.L.K. Mategga to Commissioner Veterinary Services, subject: Singo Ranching scheme, ranch number 17, in File No. C.40: Singo and Buruli Ranches, MAAIF, Entebbe.

56 Minute 6/71, in minutes of the meeting of the RAC held at Kampala on 10 September, 1971, in File No. C.120 (a) Ranching Selection Board, MAAIF, Entebbe.

57 Ref C.MAI/V/8/22 of 22 December, 1971 from Marples, Sec. RAC, to M/s Tibaimanya and Company, P.O. Box 179, Mbarara. Subject Ranch Number 6, Masaka ranching Scheme, in File No. C.120 (a), RSB, MAAIF Archives, Entebbe.

58 Minute 5/66: Selection of applicants for interviews and allocation of ranches: Minutes of 3rd meeting held at the MAIGF on 15 July, 1966, in File No. C/A3 RSB, MAAIF, Entebbe.

59 Minute 4/72, minutes of the 3rd meeting of the RAC of 31 January, 1972, in File No. C.120 (a) Ranching Selection Board, MAAIF, Entebbe.

60 Ref. SDS/51/7 of 4 October, 1971 from B.G. Semafumu, for Commissioner of Veterinary services and Animal Industry to Mr. A.K. Bukenya, P.O. Box 212 Kampala, subject: Ranch number 6, Buruli Ranching Scheme, in File SDS/51: Singo and Buruli Ranches, MAAIF, Entebbe.

61 See letter from the Chairman, Ankole Kingdom Land Board, of 8 Oct., 1963, to all saza chiefs, subject: Enclosure of Agricultural of Pastoral Lands, in File Lan 8: Land Policy – Minutes of Meetings and correspondences from Ministry Headquarters, opened Oct. 8, 1963, District Administration Archives, Mbarara.

62 See C.2/10 of 6 December, 1969 from J.H. Kagoda Commissioner VSAI to DVO, Lango, in File No. C.2, Correspondences – Maruzi Ranching Scheme, MAAIF, Entebbe.

63 See C1 of 27 November, 1969, subject – Land under Development for Ranching at Akokoro from B.B. Mayanja, in charge of Maruzi Ranching Project, to D.C. Lango, in File No. C.2, Correspondences – Maruzi Ranching Scheme, MAAIF, Entebbe.

64 See letter of 16 October, 1967 from K.K. Karyegyesa to Secretary RSB, in File No. C/A3 RSB.

65 See Hon. J.K. Babiiha, M.P., 'Opening Address by His Excellency the Vice President/ Minister of Animal Industry, game and Fisheries, Uganda,' in Sacker and Trail (1968).

66 In File C.120, opened July 1969, Ranching Advisory Committee (RAC), MAAIF, Entebbe.

67 See Hansards, issue No. 14, 28 June to 23 August, 1990, pp. 377.

68 In Ref. C.40 loose minute by B.B. Mayanja, Deputy commissioner veterinary services and animal industry of 28 October, 1978 to Commissioner Veterinary Services and Animal Industry. Subject – Trip to Singo ranching Scheme on 25 October, 1978, in File No. C.40: Singo and Buruli Ranches.

69 In Ref. C.40 loose minute by B.B. Mayanja, Deputy commissioner veterinary services and animal industry of 28 October, 1978 to Commissioner Veterinary Services and Animal Industry. Subject: Trip to Singo ranching Scheme on 25 October, 1978, in File No. C.40: Singo and Buruli Ranches.

70 See C/MUB/KBG/45 of 17 December, 1982 from Jonan Tumwebaze-Kabachelor, ADC, I.C. Kiboga sub-district, to Commissioner, VSAI. Subject: Singo ranching scheme, in File No. C.40: Singo and Buruli Ranches.

71 From Dr. A.K. Sentamu officer in charge, Singo ranching scheme, to Commissioner VSAI, of 12 September, 1983, in File No. C.40: Singo and Buruli Ranches.

72 In Ref. C.40 letter from Dr. Ananias Lubega of 25 March, 1986 to PS MAIGF. Subject: rehabilitation of Singo ranching scheme, in File No. C.40: Singo and Buruli Ranches.

73 From A.L. Iga officer in charge, Singo Ranching Scheme, to PS MAIGF. Subject: Report on ranches within Singo Ranching Scheme, in File No. C.40: Singo and Buruli Ranches.

74 Ref. C.2 of 10 August, 1989 from office of the Commissioner to Deputy Minister (Research), MAAIF, subject: A Brief on Maruzi Ranch, in File No. C.2, Correspondences – Maruzi Ranching Scheme, MAAIF, Entebbe.

75 Minutes of the meeting of ranchers in Buruli ranching scheme and the RVO, Buganda held at Nabiswera, Buruli on 7 May, 1971, in File No. C.40: Singo and Buruli Ranches. The minutes were signed by Ananias L. Iga, RVO Buganda and was attended by owners of ranches number 1A, 1B, 1C, 1D, 2A, 2B, 3A, 3B, 5A, 5B, 5C, 7B, 8A and 8B.

76 'A short review of Ranch Development and Cattle Density in Buganda Region,' by Regional Veterinary Officer, Buganda, Mr. A. Iga, presented to the meeting of RAC of 10 September, 1971, in File No. C.120 (a), RSB, MAAIF Archives, Entebbe.

77 Minute 1/71, in minutes of the meeting of the RAC held at Kampala on 10 September, 1971, in File No. C.120 (a) Ranching Selection Board, MAAIF, Entebbe.

78 Ref. BC.130/17 of 30 September, 1971 from Robert Ekinu, Secretary to the Treasury to the Commissioner for veterinary services, in File No. C.40: Singo and Buruli Ranches.

79 Ref. C.41 of 20 March, 1974 from B.B. Mayanja, Commissioner VSAI to Commissioner Lands and Survey. Subject: Compensation of Evicted Tenants, File No. C.108/C, opened June 1970: Mawogola Ranching Scheme.

80 Ref. C.40/33 of 7 February, 1972 from J.H. Kagoda Comm. Veterinary services and animal industry to Commissioner for Budget, Ministry of Finance. Subject: Compensation in Singo Ranches, in File No. C.40: Singo and Buruli Ranches.

81 Ref. BC.130/17 of 15 February, 1972 from Robert E. Ekinu, Secretary to the Treasury to Commissioner, Veterinary services and animal industry. Subject: Compensation in Singo Ranches, in File No. C.40: Singo and Buruli Ranches.

82 Ref. C.120/48 of 11 February, 1972 from Marples to PS MAIGF, in File No. C.120 (a) Ranching Selection Board, MAAIF, Entebbe.

83 Ref. C.120/48 of 11 February, 1972 from Marples to PS MAIGF, in File No. C.120 (a) Ranching Selection Board, MAAIF, Entebbe.

84 HJSM/JK of 8 February 1972 from H.J.S Marples, Commissioner Veterinary Services and Animal Industry to PS/Comm. MAIGF, in File No. C.120 (a) Ranching Selection Board, MAAIF, Entebbe.

85 From A.L. Iga officer in charge, Singo Ranching Scheme, to PS MAIGF. Subject – Report on ranches within Singo Ranching Scheme, in File No. C.40: Singo and Buruli Ranches.

86 Letter from A.M. Sibo, Bihogo ranchers of 6 December, 1971, to Commissioner Veterinary Dept., MAIGF, subject, application for a Ranch, in File No. C.120 (a), RSB, MAAIF Archives, Entebbe.

87 Letter from A.M. Sibo of 4 April, 1972 to PS Ministry of Animal Resources, in File No. C.120 (a), RSB, MAAIF Archives, Entebbe.

88 Letter from Nafutali Musajjalumbwa of 10 March, 1980 to Commissioner Veterinary Services. Subject – Application to Graze on Ranch number 4A, Kalyampande, in File No. C.40: Singo and Buruli Ranches.

89 See 'LDUs threaten Lake Mburo,' *New Vision*, April 25, 1991, pp. 8-9.

90 In *Clay* (1984), it is argued that the expulsion of especially Banyarwanda pastoralists from the Kanyarweru area starting in 1982 was partly instigated by their alleged support for the then guerrilla National Resistance Army (NRA).

91 Whereas the Act of Parliament that created LMNP was passed in July 1984, the communities, which had settled in area, were forcefully evicted following the expiry of the January 29, 1983 ultimatum. Houses were set ablaze, livestock confiscated or shot, women raped and a few people lost their lives (see GAT-Consult and Norconsult AB, 1992).

92 See 'The problem of Squatters in Ranches,' *Financial Times*, 6 August 1990, pp. 6.

93 The modernization of agriculture is central to government's economic development strategy. This is because of the recognition that the country's economy is dominated by agriculture, and remains dependent for economic growth on agricultural growth. To meet the demand for food by a growing population; to generate foreign exchange earning required to import needed agricultural inputs which the country does not produce on its own, and to improve the standard of living of the people eradicating poverty, government has had to promote growth of the agricultural sector (Republic of Uganda, 1997; MAAIF, 1996).

94 See *Parliamentary Hansards*, issue No. 14, 28 June 1990 to 23 August 1990, pp. 381.

95 See, 'Mburo in Danger,' *New Vision*, 14 February 1992.

96 This information is contained in a memorandum from squatters to the Chairman of the Bunyoro ranches Restructuring Committee, of 20 November, 1990. The information was corroborated during an interview with the then LC3 Chairman, Kiryandongo sub-county, Kibanda county, Masindi District, Jonathan Bigirwa who was one of the key people advocating for the rights of the squatters.

97 See, 'Mburo Squatters Compensated,' in *New Vision*, 22 April 1994, pp. 8.

98 See, *New Vision*, 18 May 1994, pp. 3.

99 In a report on Ranching Development in Uganda, December 1990, Ministry of Agriculture Animal Industries and Fisheries (MAAIF), Entebbe.

100 See 'Telephone Ranchers stand to lose most,' *Weekly Topic*, 24-31 August 1990.

101 See ' Ministers Visit troubled Ranches,' *New Visions*, 17 August 1990, pp. 1.

102 See ' Squatters take up arms against Ranchers,' *Financial Times*, 20 August, 1990.

103 See 'Military gets involved in Mawogola ranches' crisis,' *The Star*, 14 August, 1990.

104 *ibid.*

105 See 'Soldiers deployed to Pacify Mawogola,' *The Star*, 15 August, 1990.

106 The subject of this memo was 'Matters concerning Squatters and Ranchers,' it ended with the word 'solidarity', and was signed by a Captain Yosama Lwanyaga (see 'Soldiers deployed to Pacify Mawogola,' *The Star*, 15 August, 1990).

107 See 'Museveni Meets Masaka Ranchers,' *New Vision*, 14 August 1990, pp. 1.

108 See 'Ranchers Row to End,' *New Vision*, 15 August 1990, pp. 1.

109 See *Parliamentary Hansards*, issue No. 14, 28 June 1990 to 23 August 1990, pp. 375.

110 See *ibid.*, pp. 375.

111 See *ibid.*

112 See 'Ranches: Tinyefuza Defends Army,' *New Vision*, 24 August 1990.

113 See Memorandum from Kafureka, V.K. Chairman, Marumba Ranching Cooperative Society, of 19 April 1995, to the Minister of State for Agriculture in charge of Ranching, Anti-Nomadism, and water Development, ref. MRC/RR/R39 – RE: 'Losses on Ranch 39 – Nyabushozi during the period 1990-1994'.

114 See 'NRC Holds Special Session on Ranches,' *New Vision*, 23 August 1990.

115 This is contained in a report titled 'Property worth Millions destroyed by 'squatters,' prepared by E.K. Kawoya, an official of Masaka Livestock Foundation (MALIFA), dated 13 August 1990.

116 *ibid.*

117 *ibid.*

118 *ibid.*

119 See 'Stolen Cattle Impounded in City,' *New Vision*, 5 September 1990, pp. 12.

120 See 'Refugee talks progress', *New Vision,* 14 August 1990, pp. 1.

121 See 'Minister accused of Manipulating President,' *The Star,* 24 August 1990, and 'Ranches: Tinyefuza Defends Army,' *New Vision,* 24 August 1990.

122 See *Parliamentary Hansards*, issue No. 14, 28 June 1990 to 23 August 1990, pp. 390-1.

123 See, '62 Ranches to be divided and Sold,' *New Vision,* 26 January, 1990.

124 This policy statement was made as a directive by President Yoweri Museveni to Minister of Animal Industry and Fisheries, Kampala (Ref. PO/5, dated 6/8/1990, in File Lan/10, Vol. 4, Land Matters, Office of the Chief Administrative Officer, Luwero).

125 See 'Government to repossess, restructure ranches,' *New Vision*, 24 August 1990.

126 The notice was published in the Uganda gazette of 12 October 1990 Vol. LXXXTTT, No. 42.

127 The decision by government (through the National Resistance Council) to establish the RRB was published in the Uganda Gazette of 13 October 1990.

128 See 'President Museveni Directs on Ranches,' *New Vision*, 21 November 1990.

129 In February 1992, President Museveni told two public rallies in Masaka District that government would soon come up with a law prohibiting nomadism because it leads to the spread of cattle diseases, and leads to overstocking and overgrazing. See 'Nomadism will be outlawed – Museveni,' in *New Vision,* 18 February 1992). During the 1992 May-day address in Mbarara, President Museveni warned that: '... government will not tolerate the practice of nomadism anymore ... those who will be found continuing with the habit after the demarcation of the government ranches, will be arrested and prosecuted ...' (See 'Museveni Warns Nomads,' *New Vision,* 4 May, 1992).

130 See 'Government to repossess, restructure ranches,' *New Vision*, 24 August 1990.

131 See 'Ranches Board given Deadline,' *New Vision*, 14 January 1997.

132 Instruction to ensure this directive was implemented within a week was given to Chairman of Ranches Restructuring Board, the District Administrators, Mbarara, Rakai and Masaka, and the concerned security organisations. The affected cattle keeper who would not comply were threatened with arrest (See 'President Museveni Directs on Ranches,' *New Vision*, 21 November 1990).

133 This was provided for in Section 4, sub-section 2 of the terms of reference of the Board, also provided for in general notice No. 182 of 1990 section 4, sub-section (i) and (h).

134 It was in the interest of squatters to declare all the animals they had at their disposal, because it was explicitly stated by the RRB at the beginning of the restructuring exercise that amount of land to be allocated would depend on number of cattle owned. It is therefore, assumed in this study that amount of land allocated corresponded to the number of animals owned by the squatters. Even during the fieldwork, it was not easy to establish the actual numbers of heads of cattle owned by cattle keepers. Traditionally, pastoralists are never keen to say the actual number of their herds. That is why we

used this method to estimate the number of animals owned by each squatter on the ranches at the time of ranches restructuring.

135 See 'Byanyima Wants IPFC to boycott May Elections,' *Monitor*, 25-27 March 1996.

135 See Republic of Uganda/Parliament of Uganda. *Report of the Select Committee on the Ministry of Agriculture, Animal Industry and Fisheries*, Parliament of Uganda, Parliament Buildings, Kampala, Uganda, March 1999, pp. 4.

137 *ibid.*, pp. 12-14.

138 See 'Government to repossess, restructure ranches,' *New Vision*, 24 August 1990

139 See 'Ranches Board given Deadline,' *New Vision*, 14 January 1997.

140 President Museveni, during the debate on the ranches on 22 August 1999, see *Parliamentary Hansards,* Issue No. 14, 28 June to 23 August, 1990, pp. 405.

141 *ibid.*

Chapter three

Local Pastoral Institutions: The Case of Karamoja

3.1 A Background

One of the core concerns of human ecology is the ways in which human populations organise in order to sustain themselves in a given environment. An attempt is made in this chapter to understand how the policy environment affects the action of cattle keeping communities in the semi-arid region of Karamoja, and how in their turn the cattle keepers influence their environment, and in the process affect each other. This chapter looks at the various types of processes that are relevant to understanding the adaptive strategies of cattle keepers, using pastoral institutions as the focus. Studies on human adaptability centre on understanding how humans interact with their environments. They seek to understand the combinations of available natural resources, technology, economy and social structure – all of which enable man to survive in a particular habitat (Bates and Lees, ed., 1996; Hawley, 1986). These adaptive processes depend on the ability of human beings to effectively resolve the problems they face on a daily basis as they struggle to survive. So on one hand, the environment is seen as a potential for exploitation by man, and on the other as a potential for the shaping of culture. However, it is important to note that unlike other living organisms, human beings are not constrained by the locally available phenomena only. The interaction between the local peoples with phenomena external to their locality enables them not to rely on locaL knowledge only. This is because today, local groups are not isolated but are part of wider systems of nation states and economies and are also affected by international politics. Manger (1996: 31) convincingly argues that it is not enough to concentrate on the local population as a unit of study but rather one should also show how supra-local processes operate.

The main difference between human adaptive strategies and those of other life forms is that humans do not only adapt to natural environments but also to social and cultural contexts. One way we achieve this is through the use of language – which is crucial for the formation of management units that plan and make decisions on the best options for their group. It is through the use of language that humans are able to transmit knowledge about the natural and social environments from one generation to another. In other words we are dependent on knowledge we have developed about our environment. To quote Sapir (1939, quoted in Warren: 3):

> ... the real world is to a large extent unconsciously built up on the language habits of the group. ... We see and hear and otherwise experience very largely as we do because the language habits of our community predispose certain choices of interpretation.

Language enables a people to perceive their social, symbolic and physical world differently from others. It is through language that a community portrays what it perceives to be real or unreal about the world out there. Central to all this is the need for survival. It is the various adaptive strategies that enable man to extract energy that is required for survival and for reproduction from the environment using the best alternative that is referred to as 'indigenous knowledge'.

The recent past has seen an upsurge of focus on indigenous knowledge – sometimes also referred to as 'traditional knowledge'. This refers to knowledge acquired by local peoples over a period of hundreds of years of trial-and-error of practices and experiences as they interact directly with their local environment. It includes the development of practices and tools that are crucial for the survival of the community like modes of hunting, fishing, trapping, livestock management, cultivation, and knowledge of local plants (Inglis, ed, 1993). This kind of local knowledge plays an important role in the regulation and balance of exploitative pressures that permit an ecosystem to maintain some form of stability. Routine ways of doing things gradually establish themselves as the customary way that things are done because they will have proven to be the best alternative for achieving specific objectives. This is often achieved after various modes and practices have been tried as people face challenges presented by both their social and physical environment, after which what proves the most efficient is treated as normal practice. It is this interplay that has enabled people to live in relative harmony with their environments with continuity in resource use practices that have withstood the test of time.

As they interact with their environment, human beings engage in divisions of labour either based on age, gender, or both. They create systems of inequality resulting in perpetuated subordination of some sections of society like caste and class. It is through the different forms of social organisation that human beings enhance their adaptation. However, social organisation does not just keep a society in balance with its resources – which is one of the main tenets of adaptation – but is rather more integrated and dynamic and wider in scope. Local management units are responsible for making decisions for the survival of their community within their own environment. But they are also involved in larger systems, which means that their role in the process of adaptation is not constrained by only locally available energy. Factors external to their locality, like trade, national and regional politics, communication, technology, and global environmental factors like global warming also influence them.

These processes are fostered by institutions that may be formal (following rules that human beings devise), or informal (like conventions and codes of behaviour). Institutions can be either created – such as through the use of constitutions that stipulate the dos and don'ts, or may just evolve over time. Local or traditional institutions are used here to mean those institutions that have evolved as individuals interact and which are meant to shape human interaction. The concept is commonly used to refer to the constraints that we as human beings impose on ourselves. Institutions provide structure to everyday life, the framework within which human interaction takes place. North (1990) convincingly puts it that when we wish to sit down for a meal, bury our dead, greet someone, etc. we know – or we can learn easily – how to perform these tasks in conformity with the set codes.

Knudsen (1995) on the other hand defines local institutions by denoting some of the characteristics which include; a) govern resource use in a bounded or restricted area; b) are devised and enforced by social groups; are informal, that is unrecognised by central authorities and not part of statutory law. He argues that the other approach takes a normative perspective, arguing that institutions shape people's actions and preferences. In other words, institutions provide a framework within which human interaction takes place; they reduce uncertainty in society by establishing a stable structure to human interaction.

3.2 The State and Pastoralism in Karamoja: A Retrospect

The state in Africa has received criticism from especially scholars for the manner in which it treats pastoral communities within their borders (Markakis, 1993; Mamdani, 1996). Even in a country like Somalia where pastoralists are in the majority, they continue to be marginalised by the state (Doornbos, 1993: 100). The obsession with what is referred to as the 'cultivation culture' by policy makers (and therefore sedentarisation) as the best mode of land use has largely been responsible for this. The pastoral way of life has not been regarded positively by most policy makers, and being a minority in most countries, pastoralists are regarded as second class citizens because of their peculiar way of life. Their mobile way of life is seen as backward and destructive, and so most government policies are aimed at changing the pastoral way of life – and this, using the models considered appropriate from the outside. Their life style is also considered incompatible with the civilised way of life (UNSO/UNDP, 1993).

The situation in Uganda has been no exception. The literature that looks at the impact of colonialism on the livelihood of the Karimojong has a consensus that the results of this intervention were largely disastrous to the Karimojong – mainly because of the attitude(s) that governed the policies.[1] The available information leads one to the conclusion that external intervention – which starts with colonialism – has had major influence on the direction that the development of Karamoja as a region has followed. These policies are therefore explored in some detail.

Even after Uganda was declared a British protectorate in 1894, Karamoja as a region long remained un-administered by the British. The only interaction Karamoja had with the outside world was with hunters and traders whose interests were mainly ivory (Barber, 1964). Over a period of many years, these traders established relations with the Karimojong and were the first major external influence on Karimojong cultures and traditions. The first major contact by the British was made in 1897 under the command of Colonel J.R.L. Macdonald whose relief column was travelling to the Sudan – through Karamoja. The team bartered for farm animals and food, and this was to remain the only official contact for another 15 years (Welch, *op cit.*). Austin's account of his experience in the expedition of Major Macdonald of the un-administered areas of Sudan, Abyssinia, and East Africa, which took them through Karamoja region, shows the heavy-handedness with which the Karimojong were treated.[2] For instance, a member of part of the column that went through Nyakwai territory of Karamoja, one Captain Kirkpatrick, was murdered by the locals for allegedly being involving with Nyakwai women. In retaliation, Major Macdonald ordered the burning of 13 villages with their granaries, and the destruction of all standing crops.

This was to be only an indication of the woes that were to follow, and were to remain the experiences the Karimojong had with their colonisers for the next 15 years – which certainly played a great role in shaping the attitudes the locals developed towards the colonial government.

3.3 The Proliferation of Firearms in Karamoja

Whereas there were no economic reasons for the colonialists to be in Karamoja, non-official interest in trade in ivory by Arab, Greek, Abyssinian,

Swahili, and some British ivory traders continued to flourish. Over time Abyssinian traders established themselves in the region and large caravans were camped in what is now known as Dodoth county and were trading in both ivory and slaves (Welch *op cit.*: 51). In the mean time, the proliferation of modern arms in the area also continued, and these arms were finding their way even into other areas of the protectorate. This prompted the District Commissioner Nimule, responsible for Acholi, to write to the Governor in charge of the Protectorate in July 1910 that two Acholi chiefs had already armed their followers with 1,200 rifles received from various traders via Karamoja (Welch, *op cit.*: 49; Barber, *op cit.*: 16). This posed a threat to the colonial government.

While looking at the history of the colonial period in Karamoja, Barber observed that:

> On the strength of their reports, border officials argued that action had to be taken, not because administrative expansion was profitable, not because there might be untapped resources, but because in military terms, the British could no longer ignore the North.[3]

On the other hand, as a result of an outbreak of rinderpest that decimated most Karimojong herds during the 1890's, the Karimojong demanded ivory in exchange for cattle from these traders as well as the right to pass through their territories. This was a method of restocking – as opposed to trinkets which they previously accepted as sufficient payment. They also received arms and ammunition with which they raided some of their neighbours, especially the Turkana (Welch, *op cit.*: 47). Because of the reduction in the numbers of elephants in the region, the value of ivory increased and traders increasingly offered to exchange firearms for ivory (Barber, *op cit.*: 16).The Karimojong needed the arms for acquiring stock through raiding, and it is also argued that the traders themselves got involved in some of these raids in order to appease the Karimojong thereby increasing the scale and intensity of the raids.[4] Barber observes that while all this was going on, the government had no detailed knowledge of the district, or of the firearms trade in particular (*ibid.*: 16). The stand of the Protectorate government was that both the human and economic cost of administering this remote region was too great. What comes out as a clear manifestation of the interest was stated by Lord Harcourt, Secretary of State, that:

> ... it appears to me both dangerous and unremunerative for the Governor of Uganda to undertake the administration of a country which is not easy to access from headquarters and which has no great resources.[5]

Major Macdonald had recommended that Karamoja be controlled by military means because of the fear of losing it to traders. The visits to the region by District Superintendent of Police P.S.H. Tanner later in 1910, and administrative officer T. Grant in 1911 resulted in the confirmation that Karamoja was a lawless and militarised area (Barber, *ibid.*: 19). It was then that the colonial government thought of reconsidering Macdonald's advice, viz., control by military means. This was effected after receiving permission from the Secretary of State for Colonial Affairs for limited positive action to control 'the North', as the region was referred to. Soon after assuming office as Governor of the Protectorate in 1911, Frederick Jackson saw that it was the traders who were to blame for the sorry state of affairs in Karamoja. He decided to close the district to all traders, allowing only one opening at Mbale, and with just occasional patrols in the area.[6]

By 1912, a permanent government northern garrison had been established – which embarked on a pacification for Karamoja through shooting pastoralists, burning their huts and seizing their livestock (Welch *op cit.*: 54). In 1916, two police posts were established in the area under Turpin, one staffed by British officers who were meant to work with the local chiefs and who had earlier in 1911 been introduced in the area (Dyson-Hudson, *op cit.*). It is apparent that the military were there only to ensure law and order, and had no government plans for developing the region.

So for more than a decade after Uganda was brought under British rule, the colonial government had not yet formally entered Karamoja region. Because it was a semi-arid region, it was not agriculturally attractive. At the time, – as was alluded to by the Secretary of State (Barber, *op cit.*) – it was usual to encourage the colonies to produce cash crops like cotton and coffee that were meant to feed the industries at home. The result was that excuses of impracticability were being made. The administrators claimed that the costs were the reason for not moving into Karamoja. Reports of raids and counter raids in the region only confirmed the signal of 'lawlessness' that had been observed the region. However, it offered the attraction of the lucrative supply of ivory, and Barber (*ibid.*) gives an account of how the Abyssinian (Ethiopian), Greek and the Arab traders had traded guns for ivory with the Karimojong. A famous British hunter, 'Karamoja' Bell, was involved in this trade and it is apparent that the competition for ivory was another factor that prompted the colonial government to use military force to exclude other traders from the region. Eventually Karamoja was brought under British rule together with the rest of the colony (See Barber, *ibid.*).

Control on the movement of the herds was in force during the period that the region was closed. Two major objectives were to be achieved: (i) to simplify the administration of the Karimojong. Cattle keepers had to settle down. The administration would then be able to control the trade in ivory that had escalated into gun trafficking, and (ii) to enforce compulsory labour regulations. It would be made compulsory for the local population to work, especially on the roads in the region. It was also during this period that the colonial government effected forced relocation of whole communities from where they had formally settled to their current locations – apparently to affect the above objectives. This period is vividly remembered by the elderly and is locally called *lokwijukua,* literally meaning time of pushing; in other words, the time when they were pushed out of their original areas of residence. This happened when two of the elders from Rupa, who were my hosts during this study, were still young boys. They could recall that this is when they came to Rupa from somewhere in the present Pian County.

Other than the establishment of permanent police posts in 1916, a political structure was also instituted which introduced a hierarchy of chiefs. The system of the southern Buganda region was adopted for the Karamoja region. This was meant to ease the administration of theKaramoja region. But Dyson-Hudson (*op cit.*) convincingly argues that this system of government did not (immediately) work as it was intended to because the Karimojong hierarchy was determined through the age system. It conferred authority of leadership of the community on the elders, and did not follow the pyramid type of administration that the system of chiefs introduced. The introduction of this new political system was the first external influence of the political organisation of the Karimojong society. The elders saw themselves marginalised and 'dethroned' by such a system and so disassociated themselves from it and the result was resistance since the people tended to listen to their leaders.

Caught in this impasse, the colonial government sought allegiance from the younger generation that belonged to the junior generation set. Some of them eagerly accepted it – probably to get out of the control of the elders and therefore make a break with the traditional system. Using the 1918 Native Authorities Ordinance, the chiefs had been given a wide range of powers. By 1919, the chiefs had managed to force into labour 40% of the adult male population of the district.[7]

These measures were met with hostile response from most of the Karimojong because they affected the very existence of their mobility and labour to take care of the herds. Part of the labour fource that was supposed to take the herds to the dry season grazing areas was required to stay behind. The colonialists soon learnt that the young men they had recruited as chiefs had no leverage in the society and thus proved ineffective in effecting government policy.[8] The apparent failure of the system justified even more state brutality towards the Karimojong as it was being forced to work.

The restriction of movement as a result of the closure of the region brought about a concentration of the herds in the central belt where the settlements were, with disastrous overgrazing, resulting in a steady destruction of the soils through various types of erosion. The blame for this ecological disaster was later to be put on the 'large herds' of the Karimojong and confiscation and forced sales of livestock followed. In relation to the destruction of the environment in the region, Baker (1975: 193), points out that '... attention was turned not to the fundamental cause of the change, i.e., the disruption of a well-tried traditional system, but to its control as a matter of soil-conservation policy ...'. The restrictions also aggravated the spread of cattle diseases due to the concentration of the animals. All the resultant situations were blamed on the 'large herds' thereby justifying de-stocking policies that were forcefully carried out.

It was after the appointment of the first District Commissioner for Karamoja in 1921 that boundaries in the region were drawn. However, when investigations to draw these boundaries were carried out, they were done during the wet season when most of the herds are concentrated in the central belt. The investigations did not appreciate the transhumane nature of the Karimojong, resulting into a misjudgement that the vast grazing land reserved for the dry season was unused land and was therefore allocated to the neighbouring tribes (Baker *ibid.*: 192). For instance, the south-western plains which had been grazed by the Bokora and Pian communities were given to Teso (Usuk area now in Katakwi district), and the dry season areas used by the Matheniko in Upe county in the south-east were given to the pastoral Pokot of Kenya. The consequences were heightened hostilities in these border areas, which have continued to date. For instance the Pokot were the first victims of the 1979 armament of the Pian, and were forced to flee the land they had settled in Upe county to join the Pokot of Kenya.[9] This was a spontaneous reaction to the interruption of a system that ensured the livelihood of the Karimojong people because they were no longer able to freely track water and pasture for their herds in times of hardship.

The marking of the boundaries around the region included the marking of internal boundaries. This, as elsewhere in the country, was done along tribes or clans within the region and these were to become counties, thus allocating land to particular groups. Whereas the Karimojong local system of herding was characterised by sharing resources, the government demanded that the herders get written permission from the local chiefs to move from one area to another. In the end, these boundaries interfered with the seasonal movement of the Karimojong that enabled them to track water and pasture within their region. The reciprocal and complementary alliances that the Karimojong had developed locally were also broken by these boundaries.[10]

Some land in Karamoja was also gazetted by government as Kidepo Valley National Park, Matheniko game reserve, South Karamoja controlled hunting area, forest reserves, and so on. It is estimated that the Karimojong lost a total of about 5,000 square kilometres in all (Mamdani, *op cit*: 23). This in turn heightened the conflicts between the different groups within the region itself since it reduced the links that had existed between them (Ocan, *op cit*). Rivalry and violent conflict for the available resources that were earlier used communally intensified, and have continued to date.

The main objective of these moves were to encourage the Karimojong to adopt a sedentary system of livestock production and/or agriculture, and to the policy makers then, the reduction of the land available for a mobile production system was one way of achieving it.

3.4 Karamoja and the Post Independence State

The stand taken by post-independence governments in Africa against pastoralists has not been any different from the one that was taken by the colonial governments. On the contrary, according to Cisterino, (*op cit.*: 90), it was harsher. This approach is to blame for the crisis that Karamoja is facing today. The policy document that was meant to usher in a new perspective and chart out a way forward for Karamoja after the failure of the colonial policies, left a lot to be desired. Instead of seeking to improve the livelihood of those in the pastoral sector, it instead sought to 'change the thinking' of the Karimojong about their mode of production – as if it is only their thinking that shapes what they do. The document observed that, 'Economic development must be centred around cattle and cattle products, but it must be cattle turned into cash, and the Karimojong must be taught to think accordingly.'[11] The cattle kept by the Karimojong were seen not as the basis of survival but as commodities ... that must be turned into cash!

The Karimojong are a group that has benefited least from the attainment of independence because of the perceptions held and policies that continue to be made about pastoralists. Because the Karimojong remained largely insignificant both socially and politically, most governments have had little concern with their fate. All that has been advocated for decades is that the Karimojong should stop mobile pastoralism and adopt a sedentary mode of production. It is generally agreed that mobile pastoralism is not only backward but also destructive to the environment. While addressing a delegation of European Union ambassadors who were on a fact finding tour of Karamoja, the Minister of State for Karamoja lamented that '... Karamoja had all along been neglected and was always considered last in government developments'.[12]

As a first step towards introducing change to the Karimojong community was the strategy of disarming them. This policy, which was started during the colonial time, was later to be adamantly followed by the independence regimes. Since the Karimojong had experienced violence from the state, they embarked on a strategy of protecting themselves. The guns that were bought from the Arab and Abyssinian traders were supplemented by locally making others (locally called *amatida*). They used these guns not only for defence purposes but also for raiding. The turning point on armament in Karamoja was the 1979 overthrow of the Amin military regime. When the head of state fled, so did the army – abandoning most military barracks, of which Moroto was one. This prompted the Karimojong to break into the armoury, which they emptied virtually of all the arms and ammunition they could carry. This scenario ushered in new dimensions to the politics of Karamoja. It sparked off a series of unprecedented cattle raids that have continued to date. As a result of this new military might, they turned these guns not only at each other but also to their agro-pastoral neighbours in the districts of Kapchorwa, Mbale, Kumi, Soroti, Lira, Apac, and Kitgum whom they raided of livestock. I shall discuss this later.

The first independence government continued to maintain a military presence in the region, among other reasons, to try to contain cattle rustling. The period between 1971 and 1979 under the presidency of General Idi Amin was very brutal for the Karimojong. Having served under the colonial regime, as well as in the first independence regime, and therefore having participated in the repression of the Karimojong, Idi Amin ruled the Karimojong with a military hand that left vivid memories to most Karimojong. He not only fought cattle rustling, but also attempted to change the social lifestyle of the Karimojong radically through, for instance, forced wearing of clothes, the refusal of which was punishable by death.

The Obote II regime (1980-85) that succeeded Idi Amin was left with the problem of providing security in the region because the Karimojong had recently acquired arms. As an antidote, a peoples militia force was recruited, trained, and equipped by the government in all the neighbouring districts for the purpose of ensuring security by keeping off rustlers. During this period, there were often bloody clashes between the militia and the Karimojong. There were also accusations that during this time, the militia (mainly from

Teso) set up roadblocks where the Karimojong were put out of vehicles and later killed.[13]

The 1985 military coup that ushered in the short-lived military government led by General Tito Okello resulted in the disorganisation of this militia force. During this period, the present government (the NRM government) had also been waging guerrilla warfare since 1980. One of the strategies that the military government adopted was to recruit Karimojong warriors to bolster its weakened military strength. When the regime was overthrown a few months later, the Karimojong fled with the newly acquired arms, boosting the arsenal in the region. All this weighed favourably for the Karimojong when they later mounted raiding campaigns against their neighbours.[14]

When the NRM government took power, it disbanded the militia force and absorbed some of its combatants into the regular army. However, it did not provide an alternative stand-by force that would counter any attempts by the Karimojong to rustle cattle from their neighbours. This situation gave ground for unimpeded raids by the well-armed Karimojong on their defenceless neighbours. The same article in the Weekly Topic (February 25, 1994) attributed these raids to vengeance for the way these communities had treated them in the past when they had a militia force.

Since the country had gained independence, Karamoja has been regarded as a special case. This view resulted in the 1964 Karamoja Act that offered the region a special status where administration and development were concerned. This status was short lived, though, because in 1971, after a change of government, it was repealed by the new regime. It was only in 1987 that the NRM government considered reinstating the special status on Karamoja. Whereas it is true that the present government has exhibited concern about bringing change to the Karimojong, some of the development projects have borne frustrating results. Before then, most of the projects in the region were carried out mainly by international Non-Governmental Organisations (NGOs) whose projects are usually ad-hoc and spontaneous responses to problems of famine, disease or violence in a particular region and are not aimed at a long-term improvement of pastoral life (Wabwire, 1993). It was the establishment of the Karamoja Development Agency (KDA) by Statute 4 of 1987 that ushered in the present state-led developments for the region. The aim of establishing this agency was to have it spearhead developments in Karamoja. The functions of the agency focus on transforming the Karimojong people by diversifying and improving their mode of production, improving facilities for social services in the region and co-ordinating all developmental projects in the region. One of the KDA's functions included providing sufficient water in the region for developing agriculture and animal industry.[15]

The achievements of KDA have been controversial, and the agency has generally been viewed as a failure – mainly because of poor leadership and financial mismanagement.[16] Although the agency received a grant from the European Community, there is very little to be shown in terms of achievements. Even the few valley dams that were built by the agency have since silted.

The establishment of the Ministry of State for Karamoja on top of KDA seems to authenticate government interest in finding a solution to the problems in the region. The government also preferred this ministry to be headed by a Karimojong himself. The assumption could be that it is the Karimojong who know their situation best and can therefore chart out meaningful ways for the development of the region. The steps taken by the ministry include organising conferences, workshops and forums attended by the Karimojong themselves, elders and the elite, and government departmental heads, NGOs and political leaders within Karamoja and from the neighbouring districts. These meetings are intended to underscore the salient issues that could become preconditions for the development of the region. Paramount is the issue of insecurity, which has been discussed among Karamoja and its neighbours.

In the Karamoja Leadership Forum of 1995, the Karamoja Task Force was launched. The Task Force is supposed to work under the Ministry of State for Karamoja. This forum, which was held in Karamoja, was organised mainly because the numerous developments in Karamoja – either governmental or non-governmental, had failed to achieve their stated goals. The Task Force was therefore established to spearhead the achievement of peace and development that have eluded the region for decades. Its formation also results from the dissatisfaction with the performance of KDA.

However, other than the usual rhetoric for the development of Karamoja, the Task Force has little to show. Sometime in 1998, the Karamoja Projects Implementation Unit (KPIU) was also formed with the same objective of fostering the development of the region, but with little to show today.

What these transit development organisations do show, is that there is still a problem of effective co-ordination and implementation of the development projects that have been planned for Karamoja. Although the present government has shown the will to see the alleviation of some of the problems in Karamoja, the means to alleviate them are still elusive. KDA, Karamoja Task Force and Karamoja Projects Implementation Unit are all under the Ministry of Karamoja, but have not realised much. Some of these organisations have failed to deliver only because of squabbles over leadership and finances. It is apparent though, that there is zeal enough to bring about changes in Karamoja. However, the real problems of the region have not been clearly understood and so the solutions being offered are inappropriate.

3.5 Set-up of the Adaptive Units

Both natural conditions and other factors, some of which are external to the immediate physical environment, have an impact on the alternatives available for the survival of the Karimojong and are responsible for shaping their responses. The two main ways in which Karimojong society is organised is territorially and through age groups. It is on these two points that the following discussion is going to focus. Karimojong adaptations are best defined by using political criteria; that is why this chapter focuses on the basic forms of political organisation.

In his book, Dyson-Hudson (*op cit.*, p.104-154) details the Karimojong territoriality, specifically describing their clan system. He argues that whereas decent groups are corporate and may have structural relations, it is because they are small, numerous, and relatively impermanent that political policy seems to be inappropriate (p. 104). The nature of the demands for security today also makes organising along the clan system inappropriate. Suffice it to mention that grazing camps are organised on the basis of loyalty and respect for the abilities of the leader to offer good leadership, and protection for the stock, and not on the basis of clan or lineage. The decision which grazing camp to join is primarily based on the security for the herds.

The Karimojong *sections* (to borrow the term from Dyson-Hudson and Novelli) – locally known as *ngatekerin* – constitute the largest territorial group. Dyson-Hudson uses the term *section* to mean the area of a country occupied by the permanent settlements of its inhabitants (p. 130). It should be noted that it is not only the members of the same section who reside in their sectional land, but also others – who are not of that section – but will have been accepted into this area by the members of the section. In the same locality, different ngatekerin live together and share the same resources.[17] For this reason, it is therefore common for people from the same *ngateker* to graze together. As a result, there is a stronger bond between people of the same section than with those of one's clan. This makes more sense because it is the people with whom one shares a neighbourhood who lend a hand in times of trouble – like responding to an attack by enemies, and also with whom one shares times of happiness like the ceremonial gatherings (*akiriket*).

Concern here is not to describe the Karimojong sections and clans because this has ably been done in other studies (see Dyson-Hudson, *op cit.*; and Novelli, Bruno, *op cit.*). There are no significant changes either in form or content of these sections and clans. The Karimojong have 10 major ngatekerin:

Ngimerimong
Ngikaleeso
Ngimuno
Ngitopon
Ngimonia
Ngimosingo
Ngitome
Ngibokora
Ngipian (also called Ngimuriai)
Ngimogwos

The Matheniko refer to the Ngimogwos as 'incorporated' into the Karimojong ngatekerin because according to oral history they were not originally part of the main Karimojong sections. The Ngimogwos came from among the Turkana, from the eastern side of the Mogwos Mountains – round about where they are currently settled. The Tepeth, (also known as the mountain people) is another group that is now accepted as Karimojong. This brings in the question of what makes a Karimojong? What characteristics does this group ascribe to itself as an identity of being Karimojong?

The Tepeth as a group were distinct from the rest of the Karimojong. In the past they were known to be primarily cultivators living on top of Mount Moroto. They spoke a language different from Ngakarimojong and had different norms and beliefs from those of the Karimojong. However, over time, these differences seem to have been blunted. The changes seem to have been initiated by their acquisition of arms during the overthrow of the Amin regime. With the newly acquired weapons, the Tepeth also turned to raiding cattle. As they acquired cattle, they needed forms of organisation that would enable them to manage these stocks and offered them security. The best option for them was to ally with their immediate cattle keeping neighbours, the Matheniko, for security and access to water and pasture since they lived on Mount Moroto and would therefore need low land to graze their cattle. Today they are part of the cattle-keeping groups of Karamoja and they also take part in some Karimojong cultural practices like initiation and rainmaking ceremonies. Most importantly, they joined the herding co-operatives of their cattle-keeping neighbours. Through these alliances the Tepeth have been able to learn to cope with the demands of herding. It is because of such practices that the Karimojong have accepted them as fellow Karimojong. They have been assimilated into the Karimojong community so much so that even the Tepeth language is at the brink of being lost; most of the present generations speak only Ngakarimojong. It is only a few elders who can speak the Tepeth language. Because the Tepeth are involved in pastoral production and share fundamental cultural values held by the Karimojong, they have generally been accepted as Karimojong.

3.6 Settlement Pattern and Division of Labor

3.6.1 The Permanent Settlement

The settlement pattern of the Karimojong is profoundly influenced by the topography of the region with the attendant climatic factors. It is therefore important to describe some of the salient features of the physical set-up of the settlements and show how they influence and are influenced by the productive activities that take place therein. It is important to note that the social relations of the Karimojong society are structured along gender lines with both men and women having distinct roles and activities that they play and do in their daily cores. The main role of the man is defined by the need to raise and maintain the herd, which provides for the bulk of the food requirements, and also forms the basis for social relations in the community. In their semi-arid environment, the labour requirements for grazing are immense since the herds require being moved to look for forage and water.

The Karimojong are known to maintain two types of settlements: the permanent settlements or homesteads (*ngireria, ere* singular) where most of the family members, and mainly women, children, and the elderly, stay. The permanent settlements are generally located in the central area of the region.[18] Indeed the village of Rupa, where this research was carried out, is found in the central region, and by the time of this study settlements had been in the same location for over forty years. There are also the temporary settlements or grazing camps (*nawii*), which are set up in the grazing areas often located some distances from the ngireria, and are the dwelling places of the herds-boys and warriors. This shall be discussed in detail later.

The homestead primarily consists of sleeping houses made of mud-and-wattle walls with grass thatch roofs, and with entrances barely two feet high. One also sees a granary or two within the vicinity used for storage of foodstuff, mainly grain, and other domestic effects. There are also roofless enclosures called *etem*, constructed from wood as high as two meters or more, closely put together to form a wall. Some skins are then put over the top to provide shade inside. Its entrance is also barely two feet high. The etem serves as the living room for the head of the homestead where he sometimes meets to chat with his contemporaries. Within the homestead, one also sees kraals (*atamanawii*) where some livestock, basically milch cows used to be kept to provide milk and blood for those living there. These kraals are now scanty fences (much of the wood having been used as firewood) and with overgrown grass. Other than milch cows, some goats were kept in the homesteads and a few are still to be found today. The different homesteads in one enclosure are separated from each other by internal fencing called *alar*. Enclosing all this is a wood-and-thorn fence called *awuas*, which has a single main entry, also a couple of feet or so from the ground. It is this whole set-up that is referred to as the *manyatta*. Just outside the manyattas are the fields where various crops like sorghum, groundnuts, maize, etc. are cultivated. However, the main fields are normally some distance from the homesteads. The crops in all these fields are protected from destruction by livestock by the use of thorn-fences that are put all around them. The larger fields are some distance from the manyattas, and of late, because of the influence of non-governmental organisations active in the region like the Lutheran World Service (that has funded projects that provide training on animal traction), much of the opening of land is done using ox-drawn ploughs.

Virtually all the activities at *ere* are the responsibility of the women. They construct the houses, the granaries, the partitions in the *manyatta*, and the perimeter fence. They are also the ones involved in cultivation. However, because driving and managing oxen for ploughing demands physical energy, the participation of men is significant during the period of ploughing. And since the cattle are not kept at the homesteads, which means that they have to be brought from the *nawii*, the involvement of the men is inevitable. In spite of that, the women dominate the subsequent activities of planting, weeding and harvesting. Since there is no irrigation, the crops are left at the mercy of the weather for germination and growth, and as I was to find out, fields are sometimes sown up to 5 times before the crop finally germinates and grows to maturity. This puts strain on labour since the same activities have to done over and over again, sometimes without tangible results.

3.6.2 The Temporary Settlement

As mentioned, the *awii* (also referred to as *alomar* or *adakar*) is a temporary settlement where the livestock are kept. They are often sited within the proximity of grass and water for the stock, and most important of all, where the security of the stock is most assured. The *alomar* may remain from a few weeks to a few months in the same location depending on the availability of water, pasture and security for the stock. This is where the stock is kept through the night after the day's search for water and pasture. In the past, the grazing camps would be relocated nearer the homesteads during harvest time, but today this has been greatly minimised for fear of exposing the homesteads to raiders. Virtually all the livestock is kept at the *nawii* all year round.

The *awii* consists of roughly constructed thorn kraals, both small and large, constructed close to each other, that are meant to separate the different herds. Temporary as the settlements are, there are no houses. The warriors and herds-boys sleep in the open in an area designated for the purpose (*aperit*) – normally located close to the main entrance. To keep warm a fire is lit and it is mainly the young boys who sleep closest to it.

During periods when it is felt there is relative security, some women move in to join their spouses. In such cases, the woman constructs temporary structures akin to *etem* and it is here that she, sometimes with some of the children, sleeps. Such a husband would be one of the lucky few sometimes to sleep in a shelter. What makes women move to the *nawii* is shortage of food at the *ere*. That is why they come with the little children to an area with a volatile state of security. Sometimes women are happy to give birth at the *nawii* because they can enjoy the availability of animal products – especially milk – and believe that both the newly born and the mother could live a healthy life because of their access to food. However, a pregnant woman's presence at an advanced stages of her pregnancy, and later with a newly born baby, is usually a big risk because in the event of a major attack on the kraal, she would find it almost impossible to flee. Previously the women visited the kraals to bring grain, beer, and for sexual activity, as well as to collect mainly milk for the permanent settlements. However, because of increased insecurity, it became more common for the men instead to make overnight visits to *ere* mainly for conjugal obligations.

The main factor taken into consideration while choosing the site for the *nawii* is the security of the livestock. Secondary to that is the proximity to a watering point, and pasture. Factors relating to human comfort have minimal significance. For instance, availability of water for human use is often not considered. If there are any women staying at the *nawii* then they will have to walk long distances to collect water for domestic use. The warriors drink water during the day while herding and take a bath so that when they return to the *nawii*, all they look forward to is to take some milk and blood. Since the survival of the people heavily depends on the survival of their livestock, such decisions are the most rational.

3.6.3 The *Transition Nawii*

In the recent past, there has sprung up a third form of settlement that is located way from the *ere* but not as far away as the *nawii* – between the grazing areas and the homesteads. The Karimojong refer to this type of settlement also as *nawii*, but for purposes of clarity I am going to call it *transition nawii*. These are more semi-permanent settlements because they last up to about five years or so in the same location. In here live Most of the family members who would otherwise be staying at the *ere* live there, the elders and other members of their family. Some of the warriors themselves also spend much time here providing security to the substantial number of livestock, both large and small, that are normally kept there. Its physical set-up shares the features of both the homestead and the *nawii*. As with all Karimojong settlements, the whole set-up has a perimeter fencing made of thorn that form the outer fencing – similar to what one sees at the *nawii* except that it is neater. Inside there are some houses – mostly built of thatch only – in which the women, children and girls sleep. There are also thorn kraals inside that separate the different stock like cattle, goats and camels. Here too, the warriors and young herds-boys sleep out at *aperit*. Some of the elders sleep at *etem*, warmed by a fire.

In the past some cows were left in the homestead to provide milk for the elders, the women and the children who dwell there (Dyson-Hudson, *op cit.*; Ocan *op cit.*). The size of the herd would depend on various factors like the number of people who depend on them, the availability of pasture and water in the vicinity because such stock would be grazed on the little pasture available near the homesteads and the number of cattle owned. As the resources get depleted, most of the animals would then be sent to join the rest of the livestock in the grazing area. Such practice would ensured that there was food at the *ere* during the dry season – of course depending on the harvest of a particular year and the availability of water and pasture for the livestock.

Over the years, though, the situation has changed. With the proliferation of firearms, cattle raids have become so ferocious that in most cases whole villages can get burned down by raiders. The Karimojong then saw fit to keep away livestock from the homestead. Instead they transferred this stock to a location that would not be too far away from the homesteads in order to maintain the badly needed supply of animal products to the families with little difficulty. Yet at the same time it was not too near to the *ere* to attract raiders who might destroy the homesteads. Because of the location, a substantial number of livestock is kept at the transition *nawii*. Most of them are milking cows, plus some goats and sheep. This is also where most of the camels are found. These transition settlements are mainly the home of the elders, not only because of an assured food supply in the form of animal products, but also because it offers them the opportunity to monitor the activities in the grazing areas more closely. Because of the sizeable herds, a number of warriors live here to provide the necessary protection. What I found out indicates that these settlements seem to have proliferated over a period of ten years or so, a period that seems to coincide with the spread of automatic weapons in the area – and therefore massive raiding (Ocan, *op cit.*; Otim, 1996).

The main activities carried out here is looking after the stock. By virtue of the location there is some level of security at the transition *nawiis* and therefore it is the younger boys who undertake the tasks of looking after the small stock like goats and sheep, and also the camels. The cattle, however, are left to the older boys because they (the cattle) attract attention and can easily be raided. The boys are normally quite young – averaging 6-10 years old. Since they are too young to live at the more dangerous *nawii*, this offers them the opportunity to contribute their labour at the shielded transition *nawii*.

3.7 Distribution of Diets

It is important to make some observations about the eating habits of the Karimojong in the different settlements. The set-up of the three kinds of settlements described above, and the availability of food, both agricultural and animal products, is reflected in the diets of those living in the various settlements. At the permanent settlements, the diet comprises mainly of agricultural products like sorghum, groundnuts, maize, cassava, and cowpeas. Since the harvest is stored in granaries at the homestead, and because those who dwell there are often physically too weak to cope with the hectic life at the *nawii*, they are given priority in consuming this readily available food. However, milk is often delivered to the homesteads mainly from the transition *nawii*, but also sometimes from the *alomar* itself. Because of the distance, milk from transition *nawii* is sometimes delivered to the homesteads while those who live at the transition *nawii* receive milk from the *alomar*. This is done to ease the delivery of animal products to the homesteads by reducing the travelling distance to deliver it.

The number of people who stay at the homestead during the dry season depends on a number of key factors such as the availability of food. In times of scarcity, the majority of the population often shifts to the *nawii* and the transition *nawii* in order to take advantage of the milk and blood from the stock – plus occasional sorghum porridge.

At the *alomar* on the other hand, the diet consists almost solely of milk and blood. In the evening after the stock has been returned for the night, the young boys select an animal for bleeding. They tie a rope around its neck tightly enough to expose the main vein. Then it is shot with an arrow with a shortened blade and the blood is collected in a wooden trough (ngatuba). The bleeding of the animals (*aigum*) is normally done in the evening. After milking the cows, the blood is then mixed with milk to make a mixture known as *ecarakan*. The herdsmen then settle down around a fire and take turns drinking from the same trough. The young boys often reserve some of the blood for themselves for the next morning.

Early in the morning the young boys eat what is left of the blood (since it will have coagulated 'eating' would be a more appropriate verb). Just as the east is beginning to brighten with the rising of the sun, the young boys go to the kraal and prod the sleeping cows so that they get up. They do this because they know that as soon as the cow gets up it will urinate. They will be waiting with containers to collect the urine (see picture below). It is used to clean the milking and bleeding utensils, and also for washing hands before milking. Urine is also used to turn the milk to curd.

Otherwise, for the rest of the day, all that the herders eat is wild fruit like *ngaponga* and *ngapedur* (tamarind), roots, and leaves that they gather while performing their various tasks. Only on very rare occasions do they get agricultural products while at the *alomar*, which are normally brought by visiting spouses. Indeed for all the time spent at the *alomar* (sometimes more than several months at a stretch) they may not get the opportunity to have any other meal that comprises agricultural products.

At the transition *nawii*, the diet comprises a combination of both agricultural and animal products. Since a large part of the family stays at the transition *nawii*, some of the agricultural food is carried there, and since there is also a sizeable number of livestock their products – and especially milk – make up a large portion of the diet. Some Metheniiko own camels that are keep them at the transition *nawii*. Camels are easier to manage than cattle and it is often very young boys (probably 5-8 years of age) who look after them. Such young boys would normally still be staying with their parents, and their contribution can only be within the vicinity of where they are staying, at the transition *nawii*. Since camels also produce more milk than cattle, their proximity to the majority of the population is crucial.

At the transition *nawii*, dinner is comprised of porridge made from milk and sorghum. The younger children and their mothers usually have cooked hot meals comprising of sorghum bread and sauce made from sour milk.

Since milk is plentiful and there is also some sorghum, there was no need to bleed the animals.

It is important to note that the interaction between the Karimojong and the external world also offers them the opportunity for alternatives to the problem of food scarcity. First, through the process of exchange, the Karimojong sell their milk to the urban community and with the money earned purchase foodstuff from the market. It is true that it is done only at a small scale. Factors like low milk production of the cows, high domestic demand, and because the Karimojong use cow urine to clean their milking utensils causing the milk to acquire the smell of cow urine, contribute to keeping trade on a small scale. Most of the urban population – largely non-Karimojong – shun this milk and therefore the market is limited. But in spite of this there are still some that sells milk. But on the whole, little milk is brought to town to be sold and this is mainly because much of it is consumed domestically. Much of the milk consumed in Moroto town is ultra heat-treated (UHT) milk from Kampala.

Secondly, the Karimojong also sell some livestock, both small and large, in order to purchase foodstuffs and other essentials. However, it should be emphasised here that herd capital is perishable and must be constantly replaced for pastoral production to be viable. Whereas this takes place through the process of reproduction, to the Karimojong the alternative system is through raids. This makes the pastoral production volatile as it is constantly faced with the possibility of rapid growth and rapid loss. The other constraint to investment in animals is through slaughter or sale of animals. Consumption is not only of livestock products like milk and blood, but also direct consumption of animals or other goods obtained through the exchange of animals. However, it should be noted that the exchange of animals has to be well calculated, because if consumption through exchange or slaughter exceeds the reproduction rate, then the pastoral unit falls below viability level (McCown et al.: 299). The concern of every pastoralist will therefore be to keep the herd off-take lower than the reproduction rate. That is the reason why the Karimojong rarely dispose of their stock and even when they do, it will normally be males or unproductive females so that in the end it is only the old, weak or sick animals that are selected either for slaughter or sale. It is no wonder that the general complaint of Moroto town is that Karimojong cattle produce hard and poor beef.

3.8 Adaptive Strategies of the Karimojong

3.8.1 Forms of Social and Political Organisation: the Age System

The character, form, and tone of life of a people who devote their time and energy, and putting their lives at risk to manage their livestock, is dictated by the demands of raising the livestock and the role their herds play in their lives. The demands for knowledge in herd management – like long and hard hours of work, courage, bravery and military skill, are crucial for the maintenance of pastoral life, and in Karamoja the age system is the most important form of social organisation through which such tasks and authority are managed.

The Karimojong do not have any royalty, nobility or traditional hereditary leaders in their society. Significance is attached to the age-sets through which provision is made for each male to have his place in the social and political hierarchy at one point in his life. The age system in Karamoja binds together men who have gone through initiation together. Ideally they should belong to the same age bracket and have mutual obligations to society for the rest of their lives. The age-system may appear simple but is based on an intricate set of principles that enable political authority to be exercised. At any one time there are two generation sets in corporate existence: the senior, and the junior generation sets, each comprised of age-sets.

The age system in Karamoja works in such an intricate manner that with casual analysis it would be difficult to comprehend. As a starting point, the terminology often used by scholars to describe the Karimojong age systems may be inadequate in presenting the reality.[19] The best we can do is to come as close to the truth as possible. The terms used here only describe the social groupings, but are not indicative of the structural value of these groups in the age-system.

The principle of the age-sets is analogous to the family. In a patrilineal society like that of the Karimojong, the head of the household is normally male, and power and authority to run domestic affairs are vested in him. The set up of the homestead shows the pre-eminence the head enjoys. The same principle applies to the age-set system. The senior generation set comprises the elders, the leaders. This is a corporate office of seniority and authority that is enjoyed by the older males among the Karimojong. This authority is manifested in public gatherings like ritual ceremonies and other cultural gatherings. The junior generation-set on the other hand is composed of the younger males in society. In principle, they are the sons of the elders. They implement the decisions of the elders.

3.8.2 Age-sets and Generation-sets

As mentioned, the adult Karimojong males are divided into two age units: the senior and the junior generation-set. By the very nature of the structure, a father and son cannot belong to the same generation-set. The authority of the household head is extended beyond the household to constitute the corporate leadership of the elders and it is a mechanism of minimising direct father-son conflict for instance as a result of disobedience. The structure is such that the elders belong to a generation-set senior to that of their sons. Their sons are initiated only into the junior generation-set. Each of these sets is distinctly named and power of seniority is alternated in a cyclical manner between the two generation groups. The senior-set at some time becomes defunct if the current junior generation-set goes through the ceremony of succession to become the senior generation-set. The cyclical form that it takes gives the system sequence of stratification with the senior generation-set occupying the highest social and political rank in society.

The two alternating senior generation-sets are named *Ngimoru* (the mountains) and *Ngitukoi* (the zebras), while the corresponding junior generation-sets are *Ngingatunyo* (the lions) and *Ngingetei* (the gazelles). When the senior generation-set are the *Ngimoru*, the junior generation-set will be *Ngingetei*, and when it is *Ngitukoi* who are the senior generation-set, the junior generation-set will be *Ngingatunyo*. The present senior generation-set are the *Ngimoru* and it is, therefore, the *Ngigetei* who are the junior generation-set. It is they who will become the senior generation-set of *Ngitukoi* when change of power takes place. As can be seen in the description above, the names of the generation-sets change when they move from junior to senior position. The junior generation-set of *Ngigetei* become to *Ngitukoi* when they become the senior set, while it is the junior set of *Ngingatunyo* who become the senior generation-set of *Ngimoru*.

Only the sons of the members belonging to the senior generation-set are eligible for initiation – and to the alternate junior set. For instance, the sons of the present senior generation-set (*Ngimoru*) are initiated into the *Ngigetei* junior generation-set. When this junior set assume the position of the senior set, they will become *Ngitukoi*. Only then will their sons become eligible for initiation and they will then be initiated into the junior generation-set of *Ngingatunyo* – who will become the senior generation-set of *Ngimoru* when their turn comes, and it goes on and on in this manner. So, while *Ngimoru* are the elders it is only their sons who can get initiated into the junior generation-set (of *Ngingetei*), and when the *Ngitukoi* are the senior generation set, it is only their sons who get initiated into the junior generation-set of *Nginga-*

tunyo. Put in simpler terms, it is the grandfathers and grandsons who share generation names. It is important to remember that in this system, a man is not born into an age-set but into a generation-set.

If members of the junior generation-set have adult sons, they cannot get initiated because they are not eligible. This generation of males are not recognised at traditional and ceremonial functions, which are the occasions of playing out the status that the various members have. This generation – the grandchildren of the elders – are called *Ngimirio* (rats). They are not formally recognized by the Karimojong age-system and during the prestigious cultural functions they are relegated to the position of observers, and can take part only in providing labour to collect firewood, assisting in transporting the animal carcass to the fireplace and roasting the meat. It is only the members of the junior generation-set who qualify to serve the elders. *Ngimirio* do not even eat the meat they have been roasting until the members of the elders and the members of the junior set have been served. It is only then that *Ngimiro* will receive something to eat, and even then, while being served, their names are called out and pieces of meat are thrown to them. In general, both the senior and junior generation-sets demand respect and obedience from them.

3.8.3 Age-sets

Each of these generation-sets is constituted by a number of age-sets. The term age-set is used here to refer to a group of men who undergo initiation to a specific set or rank within a generation-set. For those who are eligible, the initiation procedure (called *asapan*) constitutes carrying out a series of initiation ceremonies that will confer to the individual, membership of a particular age-set. This is a three-part ceremony that is performed by the one undergoing initiation. The first part includes the main ceremony of killing a bull (by spearing) that is then roasted. The function takes place on the ceremonial grounds of the community and is presided over by the elders who induct those seeking initiation. Each candidate will kill an animal (preferably a bull, but if one does not have one he may kill a he-goat) for the *asapan*. The elders then perform the blessing ceremonies for all those who are being initiated before the roasting starts. The most senior elder in the congregation smears rumen on the head, chest and belly of each candidate. A prayer is then offered by the prayer leader, after which the feasting takes place. They then sit in the traditional ceremonial 'U' shaped *akiriket* arrangement and all take their places. This is the most significant part of the ceremony.

The second part takes place at the homestead. During this occasion the mother of the candidate cooks the tongue, windpipe and lungs of the sacrificed bull.[20] The elders are then invited to eat this meat and will again offer blessings to the man being initiated and his family.

The final initiation ceremony involves a feast prepared from grain and vegetables of specified kinds and eaten at the kraal in the homestead. Friends and fellow candidates of the man being initiated partake of this feast where the mother and sisters of the candidate serve them with the meal they have prepared. After this stage, the ccandidate then becomes *ekasapanan*, an initiated man.[21]

The *asapan* ceremonies are one of the uniting functions where all the sections of the Karimojong[22] are involved at the same time within their localities. After ascertaining that on the average the harvest in a particular year has been good through much of Karamoja, word is spread to all the communities to the effect that there will be *asapan* in that year. This is normally after harvest between the months of August to October immediately after harvest and before the dry season grazing starts. This is because all the initiation stages are marked by a lot of drinking of the local brew made from sorghum (which shows how important cultivation is to the Karimojong). This stage should follow a good harvest. After undergoing the whole initiation process, a candidate is then considered initiated and with his colleagues will belong to a particular age-division. This is what is called an age-division because the age-set constitutes members of five batches of initiations. It is only after the last group is initiated that the age-set is fully constituted. The process of initiation is distributed in time, and this forms the basis for seniority within an age-set. Those who get initiated first within the age-set are considered seniors to those in later age-divisions of the age-set. This hierarchy is critically observed during the ritual ceremonies.

Since the ceremonies associated with initiation require agricultural products, the frequency with which they are held depends on how good the harvest has been. As we have already seen, on average, four out of every five crops result in failure. A good harvest may be localised in a small geographical area. This means that the completing initiation of an age-set can take from five to fifteen years. The explanation given for this by the elders is that it is meant to symbolise the four teats of the cow-udder.

The initiation into the current junior generation-set has been going on for decades now. The first age-set *ngimeguro* (also referred to as *ngikangaarak*, which literally means 'those who opened') started recruiting in about 1958 immediately after the power succession that saw the senior generation-set of *Ngitukoi* hand power to the then junior generation-set of *Ngingatunyo*.

Ngingatuny then became *Ngimoru*, the elders of today. The age-sets of the *Ngingetei* comprise the following:

Ngimeguro (ngikangaarak)
Ngingoroko
Ngimerikorwa
Ngiuwa
Ngiwapeto
Ngiiru (yet to come, and are said to be the last age-set, after which there will be a hand over of power)

The current age-set recruiting is the *Ngiwapeto*. They have had three age-division initiations and have one more to go before the next age-set of *Ngiiru*. They are expected to close the initiations of the generation-set of *Ngingetei* and start recruiting them. *It* is only after this that *Ngimoru* are expected to hand over power to the *Ngingetei*, who will then become the senior generation-set of *Ngitukoi*. This will then give the opportunity to some of the *Ngimirio,* who are the sons of *Ngigetei,* to be initiated.

The initiated men are the ones who receive direct counsel and instruction and execute decisions from the elders. During *akiriket* and other traditional gatherings and ceremonies, they constitute the group that is directly responsible for the arrangements preceding such a gathering. They prepare the traditional meeting place and ensure that the fuel-wood for roasting the meat has been gathered. Whereas the honours to do all this is theirs, in practice they mobilise their sons to carry out all the activities – except serving the elders when the meal is ready. They (the junior generation-set) are the main workforce responsible for the management of the herd and the general security of both the herds and the community.

3.9 Eligibility for Initiation

Whereas in principle every adult Karimojong male is eligible for initiation, there are some fundamental conditions that determine which generation-set you are initiated into.. In short, eligibility for initiation is according to which generation-set your father belongs. It is only the sons of the current senior generation-set who are eligible for initiation. When a son comes of age, when he is considered a man capable of defending himself, which is normally at puberty, he can offer to get initiated. However, a grandson of the elders is not eligible for initiation irrespective of his physiological age. In other words, the deciding factor for who gets initiated is not just age but is rather the relationship with the reigning senior generation-set. For instance, one of the sons of the chief elder of Rupa village was probably 15 years old when he got initiated – just because he is the son of an elder. On the other hand, my guide was a 45-year-old man, who already had grandchildren himself, but was not, and could never be initiated because his father is not an elder, but a *Ngigetei*. As long as the generation set of *Ngimoru* is still in power he will remain insignificant in the traditional cultural society because he cannot be initiated. In the Karimojong age system, the physiological age is subordinate to social age as defined by the strata of the age system. In the above case for instance, the 15-year-old boy occupies a higher rank and place at *akiriket* than the 45-year-old man.

On the other hand, if a woman produces a boy outside marriage and she does not get married later, the boy will take up the generation-set of his maternal uncles. If the uncles are the sons of an elder, then such a child is eligible for initiation. This is because traditionally, among the Karimojong, a child born to an unmarried woman belongs to the maternal family, and that is why if he is a boy he takes after the identity of his uncles. However, if the mother of the boy gets married, the child will takes up the identity of the mother's husband. This is because according to Karimojong tradition, a man inherits the children of a woman after marring her. If the husband belongs to the junior generation-set, the son loses the opportunity of getting initiated until his mother's husband becomes an elder. He would be eligible only if the mother gets married to an elder.

Whereas what is described here is in agreement with what Dyson-Hudson (*op cit*) has described in general terms, Dyson-Hudson did not discuss certain details especially pertaining to the dynamics of eligibility, which are crucial to understanding the principles of the Karimojong age system. However, what is described here is contrary to what Ocan (1992: 38-39) argued when he described the labour role of the age-sets. To quote:

> ... there were 50-year olds initiating with those as young as 16 years. This means that even at 50 years of age a man may still be aspiring to enter the labouring class regardless of his age since the communal mechanism which would have guaranteed his journey through the age-sets has been transformed.
>
> The basic explanation for this scenario is that the basic means to enter an age-set, because ceremony remains part of the criterion but no longer ceremony sponsored by the community, it is increasingly becoming unaffordable to many people. Those whose parents possess a lot of cattle and grain or have acquired wealth through raids or trade necessary for the cere-

mony join the warrior set earlier and even rise to higher generic sets, by fulfilling the necessary rituals at a much younger age.

It is not just the ability to produce the cattle required for the initiation festivities to qualify for initiation, but rather the factors and conditions described above. On the other hand, a candidate does not "rise to higher generic sets" as Ocan states. Only after he has been initiated into a junior generation set that can he hope one day to climb to the generation set of elders when there is a change of leadership as happened in 1958.

Asapan is traditionally marked when the candidate surrenders his clothing to the uninitiated as a symbol of transition from boyhood to adulthood. When a candidate gives away his clothing it symbolises leaving childhood. Some individuals were reluctant to get initiated because they did not want to look for new clothing. For example, the two senior elders with whom the researcher worked closely, Apa Iliokom (chief elder) and Apemago (the second elder) are age-mates and say they grew up together. However, Apa Iliokom belongs to a more senior age-set than Apemago and is therefore his chief elder. Asked why they belong to different age-sets, Apa Iliokom explained that Apemago had avoided initiation for a long time every time it came up. He recalled that he got married and even produced children before Apemago accepted initiation. Apa Iliokom belongs to *Ngiwaria* age-set while Apemago belongs to *Ngidonyokorwa*. Between these two age-sets are three others, and since each age-set recruits four times, Apemago took a minimum of 16 years to get initiated after Apa Iliokom (ideally possible only with a good harvest every year). Apa Iliokom observed that even though during their times there was no clothing to surrender, the skin used for sleeping and other personal possessions had to be surrendered instead. This was why some people he referred to as 'lazy' were reluctant to get initiated. However, today people are eager to get initiated and look forward to it. The only handicap is if they are not eligible.

The other reason given for delayed initiation is that some individuals are not able to produce the animal to be sacrificed for the ceremony and for this reason keep postponing it until they have acquired one. Some get around this by borrowing the sacrificial animal from some close kin or friend. The senior generation-set (*Ngimoru*) have an age-set called *Ngirony*, which constitutes those who were forced to get initiated after they had postponed initiation for long for various reasons. These individuals were rounded up by their peers and whoever claimed not to have a bull to sacrifice, was lent one.

Today the other factor that contributes to delayed initiation is education. As more and more Karimojong get a formal education, they are withdrawn from the daily life of the pastoral culture. Since sometimes they are at school for a long time, they get initiated at an advanced age. This is also caused by the fact that educated youth are attracted to urban life and spend some timein towns, even after dropping out of school. They move to urban areas trying to eke out an alternative form of livelihood other than pastoralism, and in such contexts the age system is irrelevant. In such cases, those who only wish to become initiated have to work with or in their community. If a Karimojong has to get married in the traditional way, is not initiated and yet is eligible, then such an individual must become initiated.

Politicians seeking for a local mandate from their people for public office often have to start by undergoing the initiation process. One example is that of one of the Ministers from Karamoja who, after being appointed as Minister of State in charge of Karamoja, had to get initiated shortly afterwards (this was sometime in 1996) in order to be able to conduct business with his people. This is because he would not command respect among the elders and others he was to work with, and therefore his authority as a government minister would have been shunned because he had not proved his manhood through initiation. Initially he had not considered it relevant since it was not relevant in the context of his life and work although he later changed his attitude.

3.10 Akiriket

In every ceremonial function held by the Karimojong, the seating arrangement is ordered in a specific manner. It is at these gatherings that the age system is most conspicuous. The people sit in a U-shaped format, with the chief elder in attendance at the centre (base) of the U (see illustration below). The rest of the elders sit on either side in a descending order according to when they were initiated, with those who were initiated first sitting nearest to the chief elder. The members from the junior generation-set then come next to the elders also according to the order of their initiation. After them come the *Ngimirio* with an informal seating arrangement, since they are not initiated and therefore command no respect in such ceremonies. These meetings are held under specific ceremonial trees, with the chief elder sitting right at the base of the trunk. The *akiriket* (specifically the chief elder) has to face Apule River, the sacred grounds of the Karimojong.

What is considered to be the most important attribute of the elders is the belief that they are the only ones who are able to intercede with the gods (*akuj*). It is their responsibility to perform rituals that are meant to cleanse their community, call for rain, and pray for protection of both the community and the herds. Such ritual gatherings could also be for the purposes of cursing

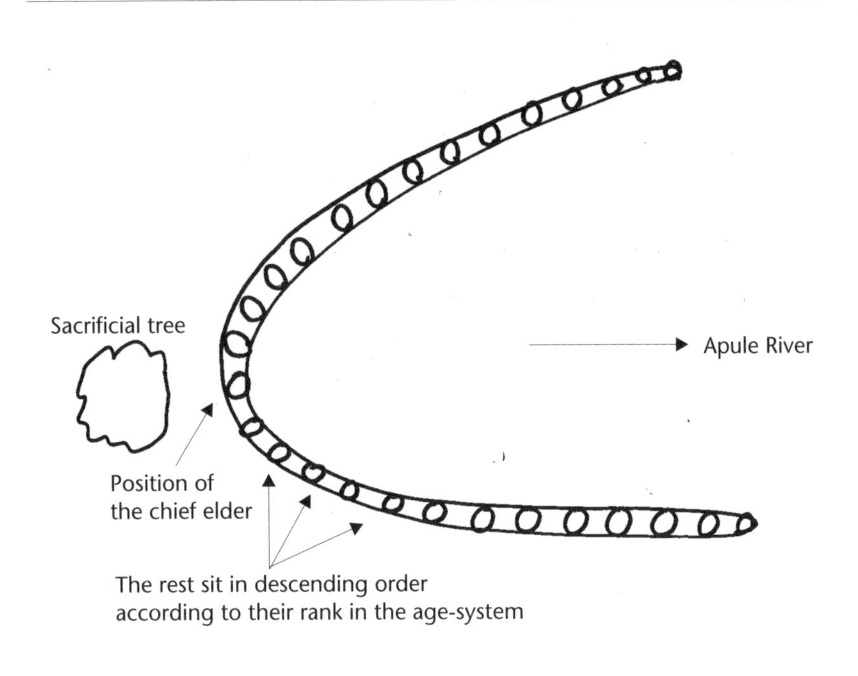

FIGURE 4 *The Sitting Order of Akiriket*

an enemy or an erring member of the community. In all these cases, the function is expected to be of benefit to the whole community.

As a result of their power to intercede with *akuj*, the elders have continued to be revered, and sometimes feared, even by rowdy warriors. The Karimojong believe that being cursed by the elders, with asured results, is lethal. They believe that if they insists on doing something without the blessing of the elders, the chances that such an act will result in disaster are high. This is why some incidents of cattle raids are looked on as minor thefts by a few individuals because they sneak away without the knowledge of others and especially of the elders, so that there will be no '*... don't do it or else ...*'.

3.11 The Dynamics of Power Relations: Authority of the Elders

As mentioned earlier, the elders hold corporate political authority and manifest this authority most significantly at cultural and ritual functions. By virtue of their seniority, society considers them wise and they are therefore relied upon to make decisions and institute policies that offer solutions to some of the challenges faced by its members. Even in everyday interaction, they are treated with deference; only their peers can challenge their opinions. This became clear to me when Apemago reacted angrily at any attempt by others to question his decisions. The others often retracted their statement quickly. It is true that as a group with shared rights, the elders have the obligation to ensure that the course of action they take is in the general interest of the community. The pastoralist mode of production demands the labour of strong and able-bodied men to traverse the vast semi-arid land with large herds of stock in search of water and pasture. Protection is needed against predators: against both mankind and wild animals. According to Dyson-Hudson (*op cit.*: 210) the elders directed the activities and through other herd owners who may not necessarily be elders, instructed both the initiated and those not initiated to implement such policies in the general interests of the community.

For generations, the Karimojong elders have wielded these corporate decisive powers, and to date the decisions on cultural and spiritual matters are still made by them. Discussions of this nature are held during the traditional gathering called *akiriket*. At these meetings the elders assert themselves as the leaders of the community. These are meetings loaded with taboos and beliefs and punitive measures for non-compliance are such that not even the warlords dare contravene its order.

Formal meetings are often called if there is an issue to be dealt with, and vary in magnitude from small kraal meetings at the *alomar,* to big gatherings that may combine more than one *alomar*. These gatherings are marked by the sacrificial killing of an animal, a he-goat or bull depending on the size of the meeting, in the traditional way of spearing. After the intestines have been removed, the 'spiritual men' (*emurwon* or fortune-teller) then spread them in a particular manner to 'read' them so that they can foretell which good fortune or tragedy lies ahead. Then the man who speared the animal (who is considered the actual giver because he pays the man who offered it) is the first to be smeared with rumen by the elders as they utter blessings to him. After the others have also been anointed and blessed, the carcass is then thrown onto a fire and is roasted whole with skin.

Before this study could be conducted, the researcher had to undergo various stages of seeking for acceptance. The researchers made a few visits to the elders' village where they explained what their mission was. This was critical since the researchers were interested in visiting the warriors at the grazing areas. This in itself was cause for suspicion, bearing in mind the volatile security situation in Karamoja today. When they were convinced that the researchers could be trusted, they welcomed them, and it was expected that they would give some valuable presents in the form of tobacco snuff and tobacco prepared for chewing. We then received blessings from the elders who anointed us with chewing tobacco as they uttered blessings as a symbol of acceptance. We were then assigned three warriors to accompany us and be our 'body guards' throughout the fieldwork period.

The first place we visited after this phase was the *transition nawii* of the most senior elder of that time, Apa Iliokom. After a few hours of chatting and many questions about what we were interested in, he decided to slaughter a goat for us as a symbol of hospitality as his guests.

After a couple of days there, we proceeded to the *nawii*. By that time the warlord in charge of the *nawii*, Nakorile, was aware of our presence in the area and that we were visiting him, and since the elders had sanctioned our visit, he had no objections. After we got there, a similar dilemma on how to feed us, as their visitors, arose. Soon after we arrived and had chatted for a while, one of the men in the group, the second in command, stood up and called the others to another shady place a few meters from where we were. What transpired in the process of getting an animal for slaughter on both occasions, gave us the opportunity to say something about the social life of the Karimojong.

3.11.1 Negotiating for the He-goat

Our presence at the *transition nawii*, had attracted the attention of most of the warriors who were staying in this region, and they soon gathered to see what these strangers wanted deep in the bush. We were received warmly since we had travelled with another elder with whom we came from the homestead. As he was our host, Apa Iliokom naturally felt obliged to feed us while we were with him. His only dilemma was that he did not have agricultural food to feed us with, and so he asked us if we could accept roast goat for our meal. We accepted. The task was then to get a he-goat that he would slaughter for us. While considering which he-goat to slaughter, he did not limit the option to his herd only because he did not himself have what he considered a suitable goat for the occasion. So he approached others for it. Since the *transition nawii* is inhabited by members of the extended family (normally an elder who is also the owner, his wives, children, his sons and their wives, grandchildren, and sometimes nephews), he had others to consult on this matter.

Apa Iliokom called over his youthful grandson, with whom he held a lengthy discussion, about lending him an animal to kill for us. Apa Iliokom started the conversation by reminding the young man that whichever animals he owned belonged to him as his grandfather. Among the Karimojong, a son does not own any stock as long as his father is alive. Even after a successful raid (which is the responsibility of the young men), the son is supposed to hand his share of the booty over to his father who then distributes it. It should not come as a surprise if the son gets the least, if he gets anything at all. This is meant to encourage the son to go for more raids with the hope that he will get more the next time. Even with the possessions that have been given to the son, the father is often free to request them, and this is exactly what Apa Iliokom was doing. He told his grandson that he had received us as his visitors and yet he does not have any food (in form of agricultural products) to cook for us, that is why he wanted to offer us a he-goat to roast for food. However, he told the young man, his own he-goats were still too young to be slaughtered and he felt sheep would not be good enough for us (the Karimojong regard killing a goat for a visitor more honourable than killing a sheep). This is the reason why he asked the young man to give him a he-goat, and promised that he would be paid back. In response, the young man argued that his goats were equally young, but the elder insisted. After a brief resistance he eventually agreed to give a he-goat for slaughter.

Similarly, when we got to the *nawii*, getting the animal slaughtered involved negotiations, except that this time negotiations lasted about two hours with everyone arguing that they did not have the right he-goat either because they were too young, or that he just had a few. Seeing what was happening, and associating it with the previous experience with Apa Iliokom, we had the impression that visitors are a burden to these people. However, it later became evident that this is what happens whenever the community wants to slaughter an animal whether it is for ceremonial or ritual purposes. This is done because individuals are reluctant to part with their stock even when they know that whatever they give will be reimbursed. Traditionally it is the person who spears the animal who actually contributes the animal and not the one who gives it. This is because he has to compensate the person who gave the animal under specified and agreed terms.

The question then is why, in spite of the fact that compensation is given – sometimes in exchange for something better – individuals are reluctant to part with their animal for slaughter? Everyone tries to show that they are not willing to part with their animals because they are so dear irrespective of whether they are compensated or not. It is common for such discussions to be held before an animal is given for a ceremony. However, the people raised the concern that in the past the discussions would not take that long before an offer was made. They argued that the change was caused because most individuals today own fewer animals, and that this has increased the attachment the individuals have to their animals. Other than that, reimbursements for the slaughtered animals often take long to be effected – also because the person who offered the animal may not readily have a replacement. There was a general disappointment because of this change, and they also observed that they used to have *akirikets* more frequently because individuals offered animals for slaughter more often, but that today they only waited for a visitor or a major ceremony.

However, after long deliberation, one of the men offered a he-goat but demanded a calf in return because the he-goat was very big. This condition prompted yet another round of negotiations and discussions on who would offer to kill it. This time it did not last as long because one of the Turkana in the kraal offered to kill it.

3.12 Mobility

The Karimojong engage in a dual economy (up to a limit the natural environment permits cultivation) cultivating food crops and rearing animals. The importance of either of these economic activities may vary from one region to another according to seasonal variations and the ability of the people to protect their herds. But there is evidence that to the Karimojong livestock are more important economically, socially, culturally, and psychologically. Livestock are the pivots around which their social, economic and cultural lives revolve. They are the source of livelihood providing the basic food (milk and blood and occasionally meat). Agricultural products like sorghum and groundnuts, if available, supplement the Karimojong's diet. Cattle are an insurance against unpredictable crop failures, and are also used for draft power in cultivation. Cattle are also a way of economic and social mobility for the men. The more cattle a man owns the more likely it is that he will be treated with deference by others even if he is not an elder. The main social occasions like marriage ceremonies, payment of bride wealth, initiation rites, and ritual activities centre on livestock and their products. In other words, cattle is virtually everything to the Karimojong, and for this reason they devote their lives, activities, and local politics towards protecting, and acquiring more cattle.

The dictates of the local environment, the contact with, and subsequently the influence of factors external to the Karimojong's locality, affect the security and prosperity of the livestock. This in turn has a strong influence on the social, political and economic organisation and behaviour of the Karimojong, as they strive to maintain their herds. The question of leadership is therefore crucial in establishing structures that enhance their ability to tap the opportunities that the local as well as the larger environment offers.

As has already been noted, Karamoja presents a natural environment with variations in soil types, rainfall patterns, vegetation cover, and availability of water. This means that the people have to develop systems that enable them to maximally tap resources from this varied environment. At the same time, Karamoja is part of the Republic of Uganda and is therefore subject to the laws of the land. Some of these laws have contributed to the cumulative crisis in Karamoja. The Karimojong have responded to their total environment by developing flexible adaptive patterns that can accommodate such variability in the natural environment, the demands of protection for their herds, and the constraints of state policies. What is at the core of these strategies is the practice of moving livestock across the different zones as the weather dictates and as far as security allows. National policies like the 'resettlement of pastoralists', as well as boundaries that do not favour mobility are also taken into consideration. The other strategies include diversification of productive activities such as cultivation, hunting and gathering, distribution of diet, and different settlement types. Focus here is on mobility.

It is common for many academicians and policy makers to dismiss local knowledge either as superstition or backwardness and they often go ahead to prescribe alternatives to particular situations with total disregard of common practices, most of the time with little success. The plight of pastoral communities is even worse. Pastoral communities have had a negative image projected on them. Concepts like *popular participation, empowerment*, and *civil society* are being used today to re-focus attention on the local and indigenous. Yet there is still little progress in empowering and allowing the participation of pastoralists in their own development (see Swift & Toulmin, 1992; in Niamir-Fuller, 1994: 12). Pastoral knowledge is still being looked at with the same static eye that accuses pastoralists of encouraging desertification, pursuing irrational principles of managing the environment, and sticking to conservative socio-cultural structures that stand on the way of development and modernisation. Such views continue to be held about a people whose adaptation is continually shaped and re-shaped by factors both within and outside of their own environment – like civil wars, the market economy, and national and regional politics.

The character and mode of life of a people who devote time and energy to herd-management is dictated by the demands for raising and protecting livestock. Pastoral production demands knowledge in herd-management, long hours of hard work, courage and military skill for protecting the stocks against predators – who may be animal or human. The Karimojong are organised around intricate institutions that have been developed over time to enable them to survive in their harsh environment. These institutions can be considered at two broad levels: the household level and community level. At household level, they include mainly local producer associations, whereas at the community level they include village leadership structures and specialised management committees (Niamir-Fuller, *ibid.*: 49). Such institutions are developed to enhance the coping mechanisms of individuals and they evolve over years of trial and error under the various physical, social, and political conditions the environment presents. When we talk about coping and survival mechanisms, these parameters define how we conceive of issues and matters around us and consequently how we apply what we know in practice to cope with such circumstances we are faced with. The institutions so developed are therefore aimed at fostering adaptation of the community by managing, controlling and co-ordinating human activities as we interact with each other and with the complex environment we find ourselves in.

The Karimojong inhabit the most productively marginal lands of Uganda and pastoralism as the main economic activity is arrived at because of the failures in alternative uses of the land with the minimum of technology that is available to them. Whereas it is true that the Karimojong practice agriculture, these efforts have most times come to naught because of the unpredictable weather. Under conditions where four out of every five crops fail (Mamdani, *op cit.* and Dyson-Hudson, *op cit.*; Cisterino, 1979: 7 puts it at every 3-4 years), the Karimojong have adopted pastoralism as the most rational system of utilising such an environment. With unpredictable and scattered rains, livestock have the advantage over crops in that they can be moved to areas where water and pasture are available.

Earlier studies among the Karimojong indicate that grazing was a tribal right, that one was free to gaze anywhere within the tribal land (Rhada and Neville Dyson-Hudson 1969: 79). However, there is little said about what alternatives the people have when they are faced with acute shortages of resources they require for maintaining their livestock. The crisis that the Karamoja region is faced with today is largely related to the destruction of the traditional mechanisms for survival that the people had developed. The flexibility that had enabled the Karimojong to survive the changing conditions of their environment were affected by the restrictions that were imposed by the state starting from the colonial era. Probably the most devastating of this was the restriction of movement through imposition of boundaries because the result was that they put stress on these areas – and the blame was instead put on the pastoralists for irrational acts like overstocking.

The practice of mobility enabled the pastoralists to move their stocks in order to escape the ecologically localised scarcity in time and space that plagues the region. The livestock must be moved to the seasonally varied grazing lands for the survival of both the herds and the people, and this means constant periodic movement in search of water and forage, and these movements are dictated by the availability of resources. The general pattern in the region is as follows. At the beginning of the rainy season, (normally between the months of April to September) herds are driven to the eastern parts of the region. There the animals can take advantage of the fresh grasses. Because of the poor rocky soils, these grasses quickly sprout with precipitation but will shrivel soon after the rains stop. It is normally after these grasses are exhausted that the herds are moved to the wetter areas to the west.

Whereas in the past this would be the time the stocks are moved to the permanent settlements to take advantage of the harvested fields, this is seldom done today because of the fear of attracting raiders to the homesteads. Today, it is mainly the cattle from the *transition nawii* that are sometimes grazed on the harvested fields while the larger *nawii* herds start their journey towards the areas often used during the dry season. When this movement starts it is common that the grasses in these areas are set on fire so that by the time the herds move in, fresh, tender, and pest-free grasses will have sprouted.[23] As the dry season intensifies, the livestock is driven to the wetter areas in the neighbouring districts. The cycle is then repeated with the return of the rains. However, at Rupa the sequence is slightly different. At the beginning of the rains, the stocks are normally moved to the rocky north-east, and as the dry season sets in the southerly movement starts targeting the more reliable watering points at Nakiloro. The reliability of the Nakiloro watering points makes the Rupa camps more rooted, as the stocks do not move out of Karamoja. However, it is appropriate to point out that the area moved to, and the duration spent in any particular area depends on security and the availability of water and pasture for the livestock, and may last from just a few weeks to months and varies from one year to another.

These movements are meticulously carried out and take place after the kraal leader decides that it is appropriate to move. The then Minister in the President's Office in charge of Karamoja Development, the Hon. David Pulkol, himself a Karimojong, observed at a workshop for the Soil and Water Conservation Society of Uganda that the movement of the Karimojong is not irrational as seen by the elite and foreigners. Instead, this movement is a scientific way of coping with the adverse natural conditions in the region. In my view it can only be called scientific in the sense that this is a tried out adaptation that enables the individuals to survive. Often before a decision to move is made, the kraal leader sends out warrior scouts to survey possible areas for relocation. Whereas water and grass are the key resources they look for, matters of security ultimately determine whether or not they move into an identified place. After getting feedback from the scouts, he then visits the sites himself to ensure that they are secure. If he is satisfied, he commands the camp to move to the new location.

3.12.1 The Grazing Camps

Grazing in Karamoja is not an individual or household exercise, but rather a collective effort where different herd owners camp together with their livestock with a common location to graze them. This practice is put in the context of grazing camps (*ngalomarin, sing. alomar*) to highlight their organisation and discuss how they function. The *ngalomarin* are probably the single most important unit that determines the survival of the Karimojong. It is not, therefore, surprising that it is also one of the units that has experienced the most significant changes as the demands for herd management change in response to various factors, internal and external, that impinge on the choices available for effective herd management. Maintenance of the herds demands political support of management units, and in Karamoja this support is mobilised and exercised locally. However, these management units have to contend with the limitations imposed on them by the Central Government in matters like land legislation, security, pastoral development s, et cetera.

As mentioned, the responsibility for the security of the livestock (and the people) has always been in the hands of the warriors. In the past the warriors were comprised predominantly by the members of the junior generation-set, and they were under direct instructions of the elders. However, today the concept invokes different connotations. The term 'warrior' no longer refers to a generation-set but rather to a physiological age group of able-bodied men capable of fighting back aggressors as they protect the herds and community. As has been indicated, there is a large group of young men today who are not initiated because they are not eligible for initiation. The irony is that they are the warriors of today, the actual fighters, the ones carrying out raids, and fighting to protect the herds and people. They are also the ones involved in acts of banditry such as highway attacks on vehicles. The organisation within the management and the protection units of the warriors deserves a closer look because of their importance for the protection of the community and herds, and the maintenance of peace and security in the region.

3.12.2 Leadership of *Ngalomarin*

Each *alomar* has a leader who in effect is its chief warrior. Although most of the warriors belong to the uninitiated generation, the leaders of the grazing camps come from the junior generation-set. They have not been elected or chosen but are individuals who have emerged as leaders as a result of proven military prowess in previous raids and wars. Such a person will have shown wit in strategic planning and a sense of good foresight on how to react to perilous situations. Since the current situation in the region is such that the chances of maintaining livestock depend on the ability to protect them against raiders, it is imperative that matters of security take precedence. The head of the *alomar* therefore presents himself as fearless and confident about his decisions and choices concerning the security of livestock. The status, dignity, and pride of Karimojong men is found in warfare and it is the desire of every able-bodied Karimojong to demonstrate as much physical and military vigour as he can. The one who excels gets following – since rational people always want to associate with winners. The choice the herd owner makes with whom to keep his herd determines his ability to lead a pastoral life because that is what determines the chances of him/her maintaining a herd, so it is treated with.

Nakorile is the commander of the camp with whom we lived all the time we were in the grazing areas. He is a mean looking and soft-spoken man, probably in his fifties, seldom with much to say. Contrary to what my expectations of a commander would be (that of a shouting and overtly cruel man) Nakorile uses reservedness to create a certain uneasiness for those around him, leaving his followers wondering all the time whether he approves of what they just did. In the end, he leaves no doubt in the eyes of a stranger, as I was, that he is in charge.

The following a kraal leader gets is the result of trust that other herd owners have put in him as capable of protecting their herds. The larger the following, the greater the honour and prestige the leader enjoys. Nakorile had over 3000 head of cattle in his kraal and there were also donkeys and goats. There was probably a 50-man strong force there, plus young boys to manage the livestock. They were the respective owners of the livestock and all were under his leadership. When the livestock is released in the morning for herding, they are not herded together as one camp but are instead split into different herds ranging anywhere from 100–400 heads. Even when they are returned in the evening they are kept in smaller individual kraals so that the owners can keep a close count of their stocks. However, the individual kraals are constructed next to each other for joint protection, only separated by thorn kraals.

Whereas the herding appears to be carried out by the individual herd owners, security, both in the kraal and in the bush during the herding, is the joint responsibility of all those who have their animals in the camp. The kraal leader is responsible for co-ordination. In this camp, the majority of the herd owners were Metheniko from Rupa, but there were also some Turkana from Kenya, and Tepeth. Whereas it is true that the Turkana have a peace agreement with the Matheniko, the reason they can afford to graze together, there seems nevertheless, to be an uneasy relationship between them.

For a long time, the Tepeth, who live on Mount Moroto, were basically cultivators. This may be because they had no land to graze cattle since there is no indication of a breed of cattle adapted to mountains. One reason they could not keep cattle was because they did not have any form of organisation or weapons to protect their livestock. However, when they acquired guns in 1979,[24] most of them turned to pastoralism since they now had the means to acquire and protect livestock. Whereas this is true, the cattle they acquired could not thrive on the mountain, so they needed the co-operation of some of their plains neighbours to gain access to water and pasture, the most vital resources necessary for them to raise their herds. On the other hand, since their traditions at the time did not involve qualities such as warrior-hood, necessary for maintaining livestock, they also needed the expertise of their more experienced neighbours. Even then, they would still need to descend to the lowlands to raise their stocks. To overcome such shortcomings, they made friends with the kraal leaders and other prominent people from the neighbouring cattle keeping groups on the lowlands including the Matheniko, Pian, Pokot, and Bokora so that they could keep their animals at their *ngalomarin*. But since the cattle-owning Tepeth are normally heavily armed, the kraal leaders readily accept them because they offer added firepower for both protection and aggression. To have Tepeth among them is also in the interest of the kraal leaders to curb the chances of their camps being raided by other groups, especially the Pokot. Tepeth have often been accused of siding with one group to raid another as they try to win favour.

The composition of the *ngalomarin* is not permanent. Individuals are free to move their stock to another *alomar* if they are not satisfied – although this is a rare occurrence. Instead what often happens is that individuals keep their livestock in more than one *alomar*. For instance, some of the people who belonged to Nakorile's *alomar* also had a part of their animals at the *alomar* of

Loodon, another prominent kraal leader in Rupa. This is done to spread risks so that if stock from one *alomar* is raided, one can turn to what is in the other *alomar*. However, spreading one's stock like this comes with a cost. It puts a heavy demand on labour because one has to contribute labour fully to all the choarsand demands of belonging to an *alomar* – some of which I will discuss below. Those herd owners who are not capable of providing the required labour to have cattle in more than one *alomar,* remedy this by allying with another herd owner in another *aloma.* He enters into a reciprocal relationship with another herd owner to whom he will entrust his animals and vice versa. This is a coping mechanism to avoid the risk of having all one's cattle raided if they are all kept in one *alomar* because of the lack of manpower to spread them.

Apa Iliokom was a wealthy man by Karimojong standards. At the *transition nawii* where he was living, he had over 100 heads of cattle, about 50 camels, and scores of goats and sheep. He also had another 2,000 heads of cattle, 600 camels, and some donkeys in two kraals. He had to distribute his stock in this manner in order to minimise the impact of loss from raids. With a large number of sons and grandsons, he did not have labour shortage.

3.12.3 Organisation of the *Alomar*

Ngalomarin are organised in a highly intricate and militarised form. As it is a large group composed of various livestock owners, internal organisation becomes crucial. We have already seen how the leader emerges and the tasks he has to undertake. But in order to run the *alomar* effectively, the leader appoints an assistant – also based on similar criteria of proven military skills – to be his second in command. The assistant helps the kraal leader plan the camp, and make decisions on the best alternatives for the safety of the stocks, some of which are critical decisions for the security of the kraal. In the absence of the kraal leader, he is responsible for planing and making the day-to-day decisions for managing the *alomar*. With the help of his assistant, the kraal leader organises and deploys the labour available to take care of various levels of defence for the herds. However, he has to give a detailed account of whatever he does in the absence of the kraal leader. The warriors do not spend all their time at the *alomar* but take turns to return home to spend some time (normally not exceeding two nights in a row) with their spouses. So the timing of who takes leave has to be properly co-ordinated in order to maintain a formidable force in the *alomar* at all times.

The *alomar* is organised in such a manner that it has patrol groups composed of warriors who take turns to patrol the area controlled by the camp on a daily basis. They are expected constantly to scout the areas that the stocks use in order to make sure that there are no intruders. They do this by meticulously checking the area for any strange footmarks – which would indicate intrusion into their territory. If they discover unknown footprints, they take the trouble to scrutinise and trace them to determine its possible origin – and therefore the likely motive of such individuals passing there. A detailed report is then given to the kraal leader in the evening and on the basis of this information, he will take a decision either to take it as a given fact or assign others – in most cases those more senior – to verify the information. If the footprints lead to another *alomar*, information is then sent to the commander of that *alomar* to protest against the intrusion and to let him know of the discovery – in a way to inform him that 'we know what you are planning'.

The routine at Nokorile's *alomar* is that after the scout groups have left to start their duty, the rest of the stocks are released for grazing in different directions. The younger boys herd most of the grazing stocks, while the older ones are involved in scouting and defence. The defence groups often leave after the stocks have been released and they track the livestock – normally up to the watering points. The different livestock is separated during grazing time. The small stock (goats, sheep) and calves normally graze nearest the camp, while the rest of the cattle is driven further. This is made easy because Nakorile's camp is located near one of the seasonal rivers called LomuNo. Like most other rivers in Karamoja, Lomuno River only has water when it has rained. During the dry season it appears as a sandy winding valley. However, under this seemingly dead valley is water, and the people know that there is water underground. Over time they have come to know exactly at which points to excavate it. At such sections, pits as deep as 4 feet with a diameter of about 2-3 feet are dug out. The depth of the pit depends on the point at which water begins to seep up from the riverbed. Someone climbs into the pit to draw the water, which is then poured into fairly large wooden canoe-like troughs (*ngatuba*) from which the animals drink. This practice is called *airu* (see picture below). The sand has to be scooped out continuously in order for the water to continue seeping out at a fairly fast rate. This is done daily until the water stops seeping out. Only small stock and small herds are watered in this manner, while the larger animals are driven to larger watering points. When this happens when the larger watering points dry up, then all the animals have to be watered at Nakiloro.

Airu is a labour intensive activity and is carried out by both the young men and girls. For the girls, this is an opportunity to prove their strength and love for livestock – which are virtues a possible suitor will be looking for in a woman. They take turns to water the animals for hours without a break and only leave after all the animals have been watered. The need for their labour explains in part the presence of girls at Nakorile's *alomar* despite the general insistence that it is dangerous for women to stay at the grazing camp. As there will be many warriors in the vicinity, this offers the girls the opportunity to act out their ability to take care of livestock. During the watering, there is often a convergence of both defence and scout groups of the warriors, who come either to participate in the watering, or to observe the activity and sometimes give instructions, as it is mainly young boys who look after these animals. After the livestock have been watered and are to be returned to the pastures, some of the defence and patrol personnel often stay behind to make sure no one takes advantage by attacking from the rear. Others will have left before the stock is driven back to the pastures – just in case someone may have laid an ambush somewhere for the returning animals.

3.13 The Question of Cattle Raids

It is true that when one talks about Karamoja today, the picture would not be complete without a discussion on cattle raids and insecurity that have characterised the region. Insecurity, and specifically cattle raiding, has often been pointed out as probably the most crucial issue affecting the social, political and economic development of Karamoja, and the ferocity with which raids are carried out keeps escalating by the day. The 1979 break-in of the Moroto army barracks revolutionised the diffusion of the gun in the region. The Karimojong and other pastoral groups in the region have continued to accumulate arms and ammunition through diverse means, but mainly from the warring southern Sudan and the Horn region of Africa. Karamoja also benefited from the collapse of the rebellion in the neighbouring Teso region between 1987 and 1990 by exchanging livestock for guns. This 'arms race' by the various Karimojong groups is prompted by the desire of each group to beat the other in military might – which is vital today for owning cattle. For instance, the Dodoth enjoyed the monopoly of military superiority in the past because they were getting weapons from the warring southern Sudan. This monopoly was, however, tipped in 1979 when the other groups also acquired guns. The result was that the Dodoth suffered heavy losses of both stock and human life in revenge attacks not only from the neighbouring Jie but also from the Turkana of Kenya. Losses were so severe that at one point in time, much of Dodoth county was without cattle and the people were trying to cling to agriculture for survival.

The camp we were living in had just moved to the present location. Nakorile left the previous location because the patrol had frequently identified footprints that, after scrutinizing them, often indicated that the Jie were spying on his territory. We learned of one incident in which some patrol warriors bumped into a group of men resting in the cool shade. While on duty, the patrol warriors had discovered strange footprints and were tracking them when they found the men. They demanded that the men identify themselves. In panic they tried to pick up their weapons. However, they were ill prepared since they had just finished drinking some milk and had put aside their guns as they were resting in the cool of the day. In such a state they were no match for the patrol warriors who were on the ready since the strange footprints had made them suspect the presence of enemies in the vicinity. As they scrambled to pick up their guns, the patrol opened fire instantly killing four of them. Two managed to escape. Upon examining the dead bodies, they confirmed that these were Jie warriors and their guess was that they had come on a survey mission for a possible raid. This was the worst encounter out of a number of cases when strange footprints were discovered. Since water from the source they were using was getting depleted, and after this incident inviting the possibility of a revenge attack by the Jie, Nakorile decided that it was time to move camp and so they had relocated further south.

The chances of being raided can be minimized but not eliminated. This is because the *ngalomarin* not only protect the stocks but also organise units for raids. The scouting parties not only look for intruders but also survey the neighbouring territories for possible attacks. Any slight lull in security can result in catastrophe. This is because every *alomar* is constantly under surveillance by others.

During one *akiriket* that was held a few weeks before we visited the *alomar*, the diviners who 'read' the intestines of the slaughtered animal had warned that the intestines indicated an imminent raid from the south-east, a likely raid by the Pokot (commonly called *Ngiupe*). To avert this raid, the community needed to sacrifice a gray he-goat. For many weeks that followed, all that people did was echo this threat. The he-goat that Apa Iliokom had slaughtered for us had had its intestines read. They had also pointed to the threat of a raid from the south-east. It seemed to remind people of the previous prediction because it sparked off a discussion in which the people blamed themselves for not treating the interpretations seriously. They kept saying it was time they carried out the sacrifice while others complained that the people who had the kind of animal that was required for the sacrifice were not willing to offer them.

A few days after this *akiriket*, the patrol warriors reported seeing some strange footprints near one of the watering points upstream of the Lomuno River. They were sure this was a group that had come to study their watering habits and promptly reported the incident to Nakorile. Sometimes raiders prefer to strike when the stocks are being watered because a number of herds come together and are therefore easier for them to drive away. This is why both the scout and patrol groups congregate at the watering areas as the animals are being watered. The enemy can only strike if they establish that they can outgun their victims, otherwise it would be suicidal. When the patrol boys traced these footprints, they suspected the intruders to have headed towards the south-east – implying that it could have been Pokot warriors. This multiplied their fears for a raid by the Pokot.

During the same period, one of the oxen that belonged to Nakorile had been stolen and information was received that it had been sighted in Tepeth area. Nakorile and his group then got involved in making frantic efforts to reach a peaceful solution by having the ox returned. A series of meetings were organised. First the matter was discussed with members of a friendly *alomar*, then with the sub-county and parish chiefs who had received a report that the ox had been sighted. However, the Tepeth denied having hidden a stolen ox. But Nakorile insisted that he had information that the ox had been transferred to Amudat, and even sent a team of 4 warriors to verify this. These meetings attracted the attention of most of the warriors because, since the ox belonged to their leader everyone wanted to show solidarity. My guess is that the enemy discovered laxity in security since most of the warriors were involved in these meetings. Apparently the warnings from the diviners, and evidence of the enemy tracking their stocks were not taken seriously, and as time would prove, they were to pay dearly for that.

At about 4:00 p.m. one afternoon, one of the herds was attacked by raiders just after the animals had been watered and were being grazed on their return journey to the camp. The herd was being looked after by 4 young boys, none of them probably beyond 13 years of age. Since most of the warriors were engrossed in the meetings, there were no patrol and defence forces to provide protection, backup power and support for the young warriors. Although all the 4 boys were armed, they were no match for the raiders. After a brief resistance that left one of them with a bullet through the thigh, the young boys realised that any attempts to continue their resistance would be suicidal and decided to flee. Over 400 heads of cattle were raided.

While this was happening, the majority of the warriors were still involved in the meetings to recover the stolen ox (see photo below that was taken at one of these meetings). They only received information about the raid long after it had taken place and it was too late to pursue the raiders. The next task was then to ascertain who carried out the raid. By the next day their suspicions were confirmed that the Pokot with the assistance of the Tepeth had done it. They even got information on where the animals had been taken. What Nakorile decided to do was to confiscate all the Tepeth herds that were in his kraal. He did this in order to put pressure on the Tepeth to bring back the cattle stolen by the Tepeth brothers. The Tepeth from Nakorile's *alomar* knew that if they did not trace the stolen cattle and return them, they were likely to lose their own cattle, which were much more than the 400 stolen. Their response was immediate, and that very evening they left the *alomar* to start the search for the stolen animals.

This is an example of how a slight lax in vigilance can result in a catastrophe. The value attached to every animal in the kraal is great, but at times probably unnecessarily high in comparison as we can see in this case. In a bid to recover one lost ox, the camp lost over 400 others, in spite of the fact that they had evidence that their *alomar* was being targeted for a possible raid. Any deviation or diversion from the vigilance that protecting the stock demands affects the ability to offer them adequate protection. The roles and responsibilities these herders play in the maintenance of the pastoralist mode of production cannot therefore be underestimated.

3.14 Politics of the Gun

The significance of the gun in the social, cultural, economic, and political life of the Karimojong today is so central that it warrants discussion. Today the *alomar* defines the social cosmos of the Karimojong. Whereas kinship ties are still strong in some cultural spheres, it is the relations that are developed with those who are your allies for the protection of the livestock at the *alomar* that are relevant in the daily social life of the warriors. Since cattle are the main source of survival to many, associates and allies in the acquisition and protection of livestock become important team-mates in the survival game. This does not mean that decent ideologies are now defunct but rather that the social constructs they invoke are undergoing a process of reconfiguration to accommodate the dynamics that the politics of the gun have introduced.

The elders of Rupa often emphasised the notion of oneness among the Karimojong groups (Bokora, Pian and Matheniko). They argued that in the past the Karimojong did not raid each other, although there were a few cases of thefts. The mass acquisition of firearms is blamed for the deteriorating relations of today. In fact, the young warriors frequently refer to the Bokora as enemies. The young generation of Karimojong does not regard other Karimojong groups as close kin and therefore could be massacred in raids. Such feelings are held in spite of the fact that Karimojong clans cut across all the 3 groups and should be the basis for cohesion. However, as mentioned before, what is more relevant is the *alomar* because it is the organisation through which acquisition and protection of livestock, the basic lifeline of the Karimojong, are effected. On the other hand, whereas the initiation ceremonies are locally held for the different groups, they are supposed to be carried out simultaneously throughout the Karimojong territory. Indeed the recent developments in the region are proof to this and put to question the supposed cohesion among the Karimojong.

Just as the *ngalomarin* are the basis of mobilisation for protection, they are also the basis of mobilisation for aggression. Large raiding parties are never from one *alomar*, but from a constellation of members from allied *ngalomarin*. These raiding parties are normally organised under the general command of one of the kraal leaders who, with the help of the other warlords, develop a strategy and then assign roles to themselves so that each attack front is meticulously planned. Raiding parties of these types and of this magnitude receive blessings from the elders, and comprise of a large force ranging anywhere from 100-500 warriors. Similarly it is common for neighbouring *ngalomarin* to ally for defence against large-scale raids. For instance, the *ngalomarin* of the Matheniko in the northern part on the border between Matheniko County and Jie county often ally to fight back the attacks from their common enemy, the Jie. Those to the north and west raid the Bokora.

Between July and September 1999, there was an unprecedented escalation in the conflicts between the Bokora and the Matheniko with a series of raids and counter raids – to the extent that it was reported that between 400-500 warriors, the majority of them Bokora, were killed in one single attack.[25] Both the Bokora and the Matheniko are fellow Karimojong and yet such loss of life could occur between them. This means the cohesion that the elders were referring to, that has created the identity of being 'Karimojong', is steadily being eroded by conflicts that create enmity, and more so as this cohesion is increasingly getting irrelevant to the young generation.

Probably the most important cultural function to the Karimojong, which involves everyone coming together to a single location is the ceremony of transfer of power from a senior generation-set to a junior-generation set. This is done at the sacred grounds at the Apule River in Matheniko County. It remains to be seen how this ceremony will be carried out since it involves all the Karimojong – who are increasingly becoming enemies.

It should be noted that local feuds are not a new phenomenon in Karamoja. What is new is the way these feuds are settled today. In the past when murder, for instance, was committed and the perpetrator apprehended, his clan would be demanded to compensate the family of the deceased with cattle. That is not what happens today. As an example, during the Christmas week of 1997, the daughter of Apemago was murdered one evening while on her way back to the village from Moroto town. The relatives of the deceased then mounted a frantic search for the murderer(s), and a few days later they had managed to trace the footprints that eventually led them to the murderer. The relatives of the accused tried in vain to negotiate for compensation because the relatives of the deceased victim demanded that the murderer be executed. In the end, the killer was led to the spot where he killed the girl and shot in execution style. This has become a common practice and even murderers arrested by government authorities are sometimes grabbed and executed by angry relatives of the deceased. The result of these kinds of revenge killings is that there will remain enmity between the affected communities, instead of appeasing or reconciling that the previous system sought to encourage. The resultant changes in the social relationships between the various sections and groups of the Karimojong can be attributed to the militarism that has engulfed this society.

Another consequence of this militarism is the distribution of ownership of cattle. The warlords and the tough warriors are increasingly amassing herds at the expense of the rest of the population. It is not surprising today to find a family that does not own any cattle. Even the warriors who participate in raids are not automatically entitled to part of the booty but will only receive a token of appreciation from the warlord if he is satisfied with their performance during the mission. This is one factor that encourages these isolated incidents of thefts that involve just a few warriors. It was reported in one local daily (*New Vision* newspaper June 29, 2000) that the Matheniko attacked the Bokora and the overall commander of the raid distributed the booty to the other commanders who participated in the raid. Most of the cattle remain in the hands of the warlord and as a result there is an emerging class of *cattle-lords* in Karamoja with the warlords and prominent warriors owning large herds while most of the population is slowly but surely getting sloughed off from the pastoral livelihood. These victims of declining pastoralism flee to seek for survival in relief food from church relief missions plus other governmental and non-governmental humanitarian programmes in the region. Others have migrated to urban areas in other districts like Soroti, Mbale, Tororo and Malaba where they have become involved in activities such as petty trade, smuggling, or begging to survive.

The influence of the market is another driving force behind the cattle raids. Most of the raiders today raid for personal gain as opposed to the traditional form of restocking for the benefit of the community. There is a frequency of major raids that occur virtually at any time of the year – as long as the warlord has planned it – and not following a major disaster as it used to be. For instance in June of 1994, the Pian raided 540 heads of cattle from the Pokot. These animals were later found on sale for slaughter at the trading centres of Namalu and Nabilatuk, and also in the neighbouring Soroti and Mbale districts.[26] The main objective of most of the raids is no longer to increase the stocks in order to increase the chances of survival in the dry season and in periods of scarcity. Instead the stocks are sold for individual financial gain, and for acquisition of better weapons and ammunition for raiding. And so the cycle continues. This was also highlighted when the State Minister in charge of Northern Uganda implicated cattle traders for fanning cattle rustling in the Karamoja region. He noted that these traders from Karamoja and the neighbouring districts have been collaborating with Karimojong warriors for the purposes of cattle trade.[27]

This is a complex network of trade that involves the exchange of livestock, guns, bullets, and money. The centre of this trade had, for a long time, been Lopedo in Kotido district, which is the first stage on the supply route for guns especially from the Sudan. A government crackdown on this market early in 1999 sent the prices of bullets soaring where they more than doubled from shillings 300 to shillings 700. Some of the prominent businessmen, the gun and ammunition dealers who used to sell and exchange guns for cattle and produce at the open market at Lopedo, were arrested. A few years earlier, access to guns and ammunition was easy because there was virtually no control by the government on this illegal bustling and lucrative trade.[28] The Monitor newspaper (January 26-29, 1996) carried a story titled 'Gun prices fall as dealers flood market'. In part it read as follows: 'In the last one year, the price of a Russian made AK-47 rifle has dropped from shillings 100,000 (US$ 100) to about shillings 50,000 (US$ 50) now in Northeast Kotido district'. This shows how easy it was for the warriors to acquire guns. These gun markets have been supplying their merchandise within the Karamoja region and beyond. An assistant minister from Trans Nzoia region in Kenya observed that the Pokot of Kenya acquired guns from the Karimojong and used them to raid cattle from their neighbours, which they then take to the Karimojong in exchange for firearms.[29]

We met a gun-trafficker on his way back from Amudat, in western Kenya, where he had wanted to sell a gun. He was driving 5 heads of cattle. Individuals carry guns on foot on a 3-day trek all the way from the Uganda-Kenya border. Most of the individuals who sell the guns are not the owners, but are only hired to go and market them. After a successful mission, they are remunerated in cattle – normally one from what they bring.

The gun has not spared the institution of marriage. Because of the prevalence of insecurity in the region, the main precondition a girl demands before she can take the hand of a man in marriage today is that the suitor owns a gun. One of the daughters of Apa Iliokom demanded to know if the researcher had a gun when he asked her whether she would marry him. The women consider the possession of guns important for their protection in the event of an attack. '... we shall be running together to look for where to hide, and (*because of that*) you can not be a man without a gun' she added. It is therefore not surprising that the desire of every young man is to own a gun because it is not only a source of acquiring and protecting livestock used as bride pride, but also as another price for a bride.

3.15 The Karimojong and their Neighbors

The sudden acquisition of guns led to the shift in the level and nature of interaction between the Karimojong and their neighbours both within and outside their region. The new might led to a frenzy of raids with devastating results. The following year, 1980, Karamoja region, and most of the country, was struck with a devastating famine. Since they had lost most of their livestock to the drought, there was need to restock, and since most of them had just acquired firearms, it was time to try out these newly acquired weapons. Large-scale raids followed in the region between the Karimojong groups and also against their agro-pastoral neighbours of the districts of Mbale, Kapchorwa, Kumi, Soroti, Lira and Kitgum.

However, following the 1979 overthrow of the Amin government, national elections were held and a new regime took office in 1980 headed by Milton Obote. Since the districts neighbouring Karamoja had offered great support to the political party that eventually took power, the Uganda People's Congress (UPC), the government felt it indebted to the people who had given it overwhelming support. Control of the Karimojong aggression was one area that this could be done. A people's militia force was then recruited, trained, and equipped by the government. This force was deployed in these regions with the main purpose of protecting local communities from the Karimojong. The years that followed saw numerous battles fought between the militia forces and the Karimojong raiding bands often with heavy casualties on both sides. For the five years this government was in power this situation continued and the animosity between the two groups went from bad to worse.

In 1986 there was a change of government in Uganda, and the new regime (the present NRM government) disbanded all the forces that existed then (which included the militia) and reorganised a national army (Museveni 1997). Members of the previous armies were to be screened before they could be incorporated into the new army. Since the districts neighbouring Karamoja had been sympathetic to the ousted regime, most combatants from these areas were hesitant in joining the new army. Instead, most started armed rebellions against the NRM government.

When the Karimojong realised that there was no more militia or any other organised-armed resistance against them, they thought it was time for revenge. At the same time the rebellion in these districts was being fuelled by some of the exiled politicians from the defeated regime by spreading the propaganda that the NRM government was punishing them for having supported the previous UPC regime. The NRM government was then faced with the task of protecting the people from these areas against the Karimojong and at the same time fighting the rebels. But since the rebellion posed a threat to state power and the Karimojong did not, the government decided to put its resources in combating the rebels. The results were devastating to these communities because the Karimojong raids could be carried out unhindered. The raids were mounted with a violence and destruction as had never before been experienced. People lost virtually all their stocks as a result of three years of relentless raids by both the Karimojong, as well as some of the rebels who were taking advantage of the situation to be able to enrich themselves (Ocan p.20). Homes and granaries were destroyed in arson attacks and livestock was taken. People lost their lives and virtually all they possessed. These raids were not only limited to the neighbouring districts but went as far as Apac, Gulu and Pallisa districts leaving a trail of destruction, killing, maiming, raping and burning of houses and granaries as they raided cattle. When there was nothing else to rustle, the Karimojong withdrew, and shortly afterwards the government brought the rebellion under control in most of these areas. The government deployed military personnel at strategic points along the boundaries with the neighbouring districts and numerous peace meetings were also held both within Karamoja and among the Karimojong and the affected communities.

Prior to the 1986 raids, the Karimojong could not cross into the neighbouring districts to graze their livestock during the dry season because of the solid military presence of the militia. They would linger at the border areas – at the risk of being attacked by the militia who feared being taken unawares. But after the militia was disbanded, this became an annual event where scores of Karimojong herds were driven to the neighbouring districts in search of water and pasture during the dry season. Since the Karimojong kept their guns, they were therefore superior in force and imposed themselves on these areas even when they were not welcome. In the past they came to these areas with their guns and numerous conflicts followed with their supposed hosts as they were accused of terrorising the defenceless people. At the end of the dry season and as they returned to their region, they raided whatever animals they would come across as the started their trek homewards.

This is a hotly contested situation with local leaders and politicians from the affected districts pressurising the government to take action to stop the Karimojong from entering their areas, or from carrying guns.[30] After exploring different options in various meetings, the parties concerned endorsed the proposal of the government to stop the Karimojong from raiding and committing other acts of violence. They should not be allowed to carry weapons when they cross into the neighbouring districts during the dry season since it is the guns that encourage raiding and other acts of violence. The argument is that the Karimojong abuse the hospitality of the neighbouring districts that allow them to graze and water their herds; they raid the livestock of their hosts as they return when the rains come.[31] In one such meeting held between the Iteso of Kumi and Soroti districts, it was agreed that the Karimojong be allowed to graze and water their cattle in Teso land as long as they were not armed. In attendance were the LC 5 Chairman for Moroto and the LC 5 Vice Chair for Kumi. The Local Defence Units from Teso, vigilante commanders, kraal leaders and elders from Karamoja were also in attendance. The meeting also accused a certain 'general' Loteng, a warlord from Kotido district, for continuing with cattle rustling.[32]

3.16 Dry Season Grazing

When the Karimojong cross into the neighbouring districts, they do not come en-mass but more according to how the resources in their grazing territories run out or how the security situation in their areas change. Since it is not likely that the resources in these areas get exhausted at the same time, the camps that exhaust theirs first will also be the first to move. But in the process, the Karimojong try to avoid as much as possible situations that may cause conflict between them. They try to keep as far apart as possible when out of their region. The kraal leaders often try to find out where the other kraals have camped and try not to get too close so that they can keep track of each other.

Previously the Karimojong could not move to these areas during the dry season because of the militia forces, and before 1979 there was little difference between the force of the Karimojong and that of their neighbours. They did not pose a serious threat. These were difficult times for the Karimojong. They used to lose many animals because of lack of water and pasture since they were kept out of dry season refuge. They consider the acquisition of guns in 1979 as an opportunity for them to force their neighbours to allow them to graze in these areas during the dry season. What they do not mention or want to accept is the fact that these guns have been used more for aggression and exclusion of other pastoral nomads from certain resources, rather than for sharing them. This argument also probably explains the daring raids the Karimojong often mounted against the well-protected areas and herds of their neighbours.

This is not to suggest that conflict between the Karimojong and their neighbours is a recent phenomenon because it has spaned decades. We have already seen that the main viability strategy for the Karimojong is mobility, and this is a transhumance movement from Karimojong territory up to the wetter areas in the neighbouring regions in the dry season. I am implying that this is a practice (plus the attendant conflicts) that must have lasted for generations. Lawrence (1957 p.10), while writing about the Iteso, refers to the aftermath of the cattle disease and famine that affected Karamoja between 1894-1896. Probably two thirds of the Karimojong perished or were scattered beyond their region. During this time there was also serious fighting between the Iteso and the Karimojong as a result of the struggle for water and grazing land. These conflicts have continued to date except that they have taken on new dimensions because of the changing social, economic, technological and political environment.

It has been shown how the territorial boundaries, both internal and national, were drawn during the colonial period. These boundaries legitimised the claims to specific territories by each group in the country. For the Karimojong, these restrictions were not feasible because Karamoja runs out of the resources necessary for the survival of the herds – water and pasture – during the dry season. They cannot manage their herds only on the resources within their territory and that is why they have to move the livestock into neighbouring regions.

Some authors have argued that the Karimojong lost precious dry season grazing land to the west during the process of demarcation because the surveys for the drawing of the boundaries were carried out in the wet season when the Karimojong were in their own territory. The vast territory used for dry season grazing was therefore considered unsettled and unused and was given to the agricultural Teso because it was considered suitable for agriculture. The implication is that even before the demarcations were made, the Karimojong did not settle on this land but were using it only during the dry season.

The practices described above are only true for the Bokora and Pian who share their western (and therefore wetter) borders with other communities. The Matheniko who are further to the Uganda- Kenya border remain in their own territory in most cases, through all the seasons. However, I should note here that they actively participated in the 1987-1991 cattle raids that were meted out against the neighbouring agro-pastoral communities.

What we see here is that the prevalence of firearms in the Karamoja region has intensified competition for resources among the Karimojong groups, as well as with their neighbours. This militant attitude has created and reproduced rivalry and enmity among previous allies like the Bokora and Matheniko. It has created feelings of military might in individuals who therefore want to exclude others from using these resources.

3.17 Taming the Gun

Because of the persistent conflicts between the Karimojong and their neighbours, one of the major tasks that various groups have been faced with in bringing peace and security to this region is controlling the use of the gun. Numerous proposals and programmes have been and are being made still, proposals and programmes that range from outright disarmament to those that aim at monitoring and controlling the use of the gun. One of these programmes deserving mention here is the vigilante programme.

The initiative that was later recognised by the government and became known as the vigilante programme started because the local political leaders were concerned about security in the area sometime in 1994. This was prompted by the deteriorating security situation in the region with cattle rustling and highway robberies on the increase. The plan included convincing the warriors to offer the control of the use of their weapons to a select group. The warriors would dedicate their guns to ensure the security of their livestock and the region as a whole, and not for aggression. This group would police the area and would apprehend the culprits and hand them over to the security organs for disciplinary action. Since the aim was to control the use of guns, the precondition to join this group was that one owned a gun.

These efforts were appreciated by the Central Government Representative (CGR) of Moroto (position of the representative of the President at the district level, now called Resident District Commissioner RDC) plus other security organs including the police and the army. The government therefore assumed the responsibility of mobilising, organising, training, and equipping the vigilante force through the national army. Since this touched upon the sensitive question of guns held by warriors, it was not accepted outright. The elders of Rupa for instance were against the idea and in one of the *akirikets* advised their sons not to be deceived by the state that had always been their enemy. They informed them that as far as they knew the government had never had good will for the Karimojong and therefore there was no reason to think that this situation was any different. The result was a withdrawal of the warriors with their guns from the villages to the grazing areas in a bid to avoid being recruited into a previously unknown institution whose motives they doubted. It was only when this happened that the government recognised the role of the elders and sent some officials to educate them and assure them that their sons would not be harmed. It took another *akiriket* to rectify the situation.

This was an incident that called for a reconsideration of the widely acclaimed notion that the institution of the elders in Karamoja had lost power to the gun-wielding warlords. The sons of most of the elders were warlords and they were certainly wary about having their sons arrested or their guns confiscated and so advised them to keep clear of government. But after they were convinced about what the vigilante force was to do, they asked their sons to register as vigilantes. The son of the second elder of Rupa, for instance, a reknown warrior is the sub-county vigilante commander for Rupa. However, the Karimojong did not fully throw their weight behind this new institution. They accepted it cautiously and only a select number of warriors registered as vigilantes while the rest remained 'regular' unregistered warriors.

In a workshop that was organised by the Centre for Basic Research in Moroto in 1994, it was clear that both the government administrators, the elders and the local leaders saw the vigilantes as the only force that could effectively check acts of violence in the region. Since those who joined the force were themselves former warriors and raiders, they were therefore familiar with the issues and the characters they are dealing with. In the years that followed, the vigilantes were instrumental in the restoration of security along the main roads in the region by tracking down and bringing to justice those responsible for shooting at vehicles. They have also been responsible for the recovery of stolen/raided cattle within the region where the recovered cattle is handed back to their original owners, sometimes with a minimum of force.

The vigilante force was put directly under the command of the army. The army provided them with basic military training and supplied them with ammunition. However, they were not stationed at the barracks. In a sense their activities were being directly monitored and under the command of the army. After their training, the vigilantes returned and continued to reside in their respective villages and *nawiis* and only came together when it was needed, for instance if a raid had been carried out and there was a recovery operation. This is also when the army would get actively involved in order to avoid any acts of revenge. The army would accompany the vigilantes and often offered backup with armoured personnel carriers (APCs) and sometimes helicopters gunships on some patrols.[33]

The vigilantes were also on the army payroll, and in 1995 for instance, the government was spending up to 60 million shillings (about US$ 60,000) in wages every month to pay for their services.[34] The commanders and members of the group have also been taken on government sponsored tours to other regions in the country to acquaint them with the progress and development made by both peasants and other pastoralists in other parts of the country.[35]

However, the vigilante force was disbanded in 1996 and reorganised into the newly created Anti Stock Theft Unit (ASTU). This is a force that was recruited not only in Karamoja but also in all the neighbouring districts with the main task of stopping cattle raids within Karamoja and among Karamoja and the neighbouring districts. It was hoped that the presence of such a force in Karamoja and all the affected areas was more effective than having only Karamoja police the warriors. In Karamoja, the vigilante force was converted to ASTU, and as in other regions, put under the control of the police force.

To date this has not taken root and there has been an escalation of violence in the region. Some of the reasons for this is that the community has not been educated and therefore the local political institutions have not approved of it. Secondly, because the former vigilantes have not understood the new structure, they have returned to their old practices of raiding.

3.18 Epilogue

Anthropologists are concerned with trying to deal with interdependencies that emerge as people interact with each other and with their environment. One area of interest is the decision-making unit, whether they are individual persons, kinship groups, et cetera. Since the environment of one individual consists of other individuals, a change in one individual's behaviour means a change in the environment of another. Human beings also do not only respond to the present but also to the anticipated future environment, and their behaviour is meant to choose the best options for the immediate goals which foster their survival, and at times taking into account the anticipated.

The forms of land use in Karamoja are closely related to the patterns of settlement and the attendant distribution of labour for the various tasks required in the production processes that take place at each settlement. The analysis in this chapter indicates that security is the central factor that determines the settlement patterns and distribution of labour. The general set-up of the *alomar* and its labour requirements has been described; how the distribution of labour is vital to the management and protection of the herds has also been shown. Similarly the set-up and activities at the homestead, and at the transition *nawii* have been highlighted to show how at each type of settlement the distribution of labour is adapted to the survival requirements and how the local politics and leadership today is meant to foster the process of adaptation.

The labour demands for herding in Karamoja are higher today than in the past when there were only two types of settlements and less sophisticated conflicts. There were fewer guns in the hands of the warriors, and labour is required on a daily basis throughout the year if the herd is to be maintained. The main concern of the management units then becomes mobilisation of adequate labour and how to manage it effectively to meet the demands for herding and protecting the livestock. This is the efficiency that each kraal leader strives for. The more the labour at the disposal of a particular *alomar* the more the chances there are that the herd capital will be maintained. This is because the available manpower will pose a threat to the competition of others and will be a deterrent to any group that might otherwise have thought of raiding it. On the other hand, such an *alomar* will have a force strong enough to mount a raid – which offers the possibility of a dramatic increase of the herd.

Over the years the co-ordination of the various roles and activities that the age system played in the community have undergone both reconfiguration and transformation, and as we have seen a number of factors are responsible for this scenario. There are presently up to three generations that have not undergone initiation into the age system although they are adult men. This has encouraged the emergence of institutions that can not only tap their labour, but also be the basis of social mobilisation for this category of individuals.

These young men have been accused of what is often referred to as theft. What normally happens is that these young men secretly plan to carry our raids and if they are successful, they keep their booty with friends elsewhere so that they may not be associated with such thefts. It is because the elders feel they have not sanctioned such raids that they insist on referring to them as thefts. In other words, they are not recognised as raids but as acts of lawlessness or banditry, and this is not only in the eyes of the traditional systems of authority but also in the eyes the government.

This situation has also been encouraged by the acquisition of modern automatic weapons that virtually every male possesses today.[36] When the elders sanctioned a raid a ritual normally preceded it to bless all those who were to participate, so that they would succeed in bringing the cattle and not meet with their death. The formation positions held by the men in the raid would also be determined by seniority in the age-sets. Those who belonged to senior age-sets would assume leadership and command of the raiding band and would be the main decision-makers during the raid.[37] Today this is the responsibility of the kraal leaders and the overall command depends on how well an individual does on the battlefield.

Highway banditry and ambushes, on the other hand, have been blamed on students –virtually all of who also belong to the uninitiated group. This is because of the exposure to a different culture of modern life and yet they cannot afford some of the material products that are associated with this modern life like shoes, watches, and good clothing, that requires cash to purchase. The alternative for them is to rob travellers. Incidents of highway robberies have been noted to coincide with the holiday period when schools are on recess. On the one hand education is seen as a means of 'developing' the people of Karamoja.[38] On the other, the educated have posed the biggest threat to development work; they are the ones who appreciate the value of material possessions. For instance, the workers of the various Development NGOs that operate in the area are robbed, sometimes at the cost of human life. The result is that some of these development agencies have had to abandon their projects because of this kind of insecurity.

The problem of the educated should be looked at from a wider context. The education system in Uganda does not offer skilled training at lower levels. Those who attempt to follow a formal school education, stop at primary or secondary school levels. They then fail to fit in the traditional lifestyle and yet, on the other hand, they do not have the necessary skills for formal employment.

Notes

1 See Dyson-Hudson *op cit.;* Baxter, P.T.W. 1975, Cisterino, 1975; Lamphear, 1976; Pazzaglia, 1982; Mamdani et al., 1992 and Ocan, 1992.

2 Austin, H.H., *With Macdonald in Uganda.* Dawsons of Pall Mall, Folkestone and London, 1903.

3 Barber, J.P., 1968, *Imperial Frontier*, East African Publishing house, quoted in Welch, *op cit.*: 50).

4 See Welch, 1969: 47; Barber, 1964: 18.

5 File 1049 Entebbe Archives. Secretary of state to Government of Uganda, 2 December 1910, in Barber, J.P., 'Karamoja in 1910.' *Uganda Journal*, 28, 1, 1964: 16.

6 See Welch, 1969: 52; Cisterino, 1979: 67. Cisterino shows how even after Uganda achieved independence there was still a notice at Iriri as one entered Karamoja that read, *You are now entering Karamoja closed district. No visitor may enter without an outlying district's permit.*

7 See Mamdani, 1996: 166.

8 Novelli, (1988: 123) quotes an extract from the report of a commission charged with drawing the Report on Security in Karamoja, commissioned by the colonial government to be given to the independent government, which confessed that: 'The chiefs cannot claim loyalty from the tribe. Their value in maintaining law and order is very limited. It follows that the strong chains of custom have been removed and replaced by the *strange* strings of local administration. We are confirmed in this opinion by overwhelming evidence we heard about the peace that obtained in the past and against the *uselessness* of the present administration. While we are all strong believers in democracy, we are the first to admit that the result of imposing it where it is neither understood nor appreciated is anarchy and chaos! This is virtually what we are faced with in Karamoja at present.' See also Dyson-Hudson, *op cit.*

9 It was only on August 9, 1995 that ceremonies were held to try to restore peace between these two groups. The 5-day ceremony was attended by elders, political and religious leaders. See *The Sunday Vision*, August 20, 1995.

10 See Dyson-Hudson, *op cit.*: 231-235.

11 See Cisterino, *op cit.*: 89, Report of the Karamoja security committee, 1961.

12 See *The New Vision*, November 15, 1995.

13 See *The Weekly Topic*, February 25, 1994.

14 In a paper he presented for a conference on peace and sustainable development for Karamoja and neighboring districts in 1994, Hon. Dan Michael Ochyengh, delegate from Kapelebyong to the Constituent Assembly, also mentions this issue.

15 See Statue 4: Karamoja Development Agency Statute, 1987, Government of Uganda.

16 See Wabwire, *op cit.*; report of the Proceedings of The Karamoja Forum, May 17-20, 1995; *The New Vision*, September 19, 1995; and *The People*, February 28, 1996. KDA is also accused of sidelining the local people in the process of project design and implementation resulting to the failure of most of the projects.

17 See also Novelli, Bruno, *Aspects of Karimojong Ethnosociology*, Verona, Kampala, 1988, p. 53-58.

18 Dyson-Hudson, (1966, p. 22-43) describes in detail the ecology of the region and shows how the varying topography and climatic conditions have resulted into different forms and patterns of utilization by the Karimojong.

19 See for instance Bernardi, 1952; The age-system of the Nilo-Hamitic Peoples, a critical evaluation.

20 Ceremonial and sacrificial animals in Karamoja are killed by spearing through the heart. The significance for this is to preserve the system of the tongue, windpipe and lungs as one. This system symbolizes continuity of life and so should be preserved as a whole and not cut. This part of the animal is not roasted during the ceremony and is instead taken home to be cooked.

21 Dyson Hudson 1966, *Karimojong Politics* (pp. 162-168) describes in detail the process of *asapan* among the Karimojong. To date, this process remains largely the same.

22 These are the Bokora, Pian and Matheniko, and in this case, even the Dodoth. the Karimojong regard the Dodoth as having been part of the Karimojong but who moved further north and were squeezed off by the Jie. The kinship terms by which they refer to the Jie – like 'brothers' strengthens this.

23 This is a contested practice with some scholars and politicians blaming it as causing environmental deterioration. However, works like that of Mamdani et al,(CBR working paper 22) argues that it is consciously done not only to provide tender grass but also to rid these tall grasses of especially ticks which cause disease to livestock. In *The New Vision* newspaper of December 22, 1995, it was reported that the burning of bush in the west in Nyakwai by the Bokora was feared to affect the unharvested sorghum in the area. However when we discussed this with the herders, they rubbished the fears saying that bush burning was carefully calculated exercise. In a more recent study I carried out for my Masters dissertation on Karimojong dry season grazing in Teso, it was clear that this was one area of conflict between the two communities. The Karimojong were accused of starting fires tat destroyed food that was both in the gardens and which was already harvested, and homes. The district administration passed a resolution that the Karimojong found to be starting these fires would be fined heavily and would also face jail terms. However, this was not easy to enforce since some of the fires would be started at night.

24 It is said that it was the Tepeth who first discovered that Moroto barracks had been abandoned by Amin's soldiers, and so they were the first to break into the armory. Their being on top of mount Moroto below which is the barracks gave them this advantage.

25 See *New Vision* newspaper, September 13, 1999.

26 *The Monitor*, June 28-July 1, 1994.

27 See the *New Vision*, May 19, 2000.

28 *The Monitor*, January 3, 1999.

29 *The Monitor* April 20, 1999.

30 See the *New Vision*, December 12, 1996, The Monitor, December 3, 1996, *New Vision*, January 17, 1997, *The Monitor* January 29, 1999 all of which report complaints by the neighboring communities against the Karimojong crossing into their areas because of fear of the plunder that most times follows.

31 Some warriors were arrested and punished by local authorities in Soroti district (*The Crusader*, January 12-19, 1996), while in another incident they were caned by for carrying guns in a market, (The *New Vision* February 21, 1996) when they were found carrying guns after this resolution was passed.

32 Reported in the *New Vision*, October 28, 1995.

33 *New Vision,* August 3, 1995; also *The Monitor*, October 13, 1995. This article also states that the idea of starting this group was championed by one Sam Abura, a retired police officer and himself a Karimojong. The group is however unpopular amongst the ardent warriors since they see the vigilantes as 'saboteurs' and are sometimes themselves targeted by these warriors. In a paper he presented at the *National Conference on Peace and Sustainable Development for Karamoja and Neighboring Districts* organised by the Ministry of State for Karamoja in 1994, the then Division Commander 3rd division also listed a number of achievements by the vigilantes within the region, which included restoration of peace and security, and hoped that if the system were consolidated even more, that cattle raiding and thuggery would reduce.

34 *The Monitor*, October 9, 1995. The paper also states that it is the 407th Brigade of the Army that effects the payments.

35 *New Vision*, November 27, 1995.

36 See also Moroto II Peace Conference: A Report of the Kenya-Uganda peace conference between the Pokot, Sabiny, and Karimojong of Uganda and the Pokot and Turkana of Kenya, 13th-16th November 1996: 7.

37 See Dyson-Hudson, *op cit.*: 173.

38 Report of the Moroto II Peace Conference: 8.

CHAPTER FOUR

Conclusions

During the colonial period the major concern for developing livestock production in Uganda was the need to increase the output of livestock products. During this period, the idea of introducing ranching schemes took shape, largely favoured by UNDP and FAO and other multi-lateral financial institutions. The colonial state justified its colonial economic development programme in Uganda through the development of commercial livestock ranching-schemes, as the case of the Ankole has suggested.

The case of the Ankole also shows that the period immediately after independence saw the policy of commercial livestock ranching crystallised. Between 1962 and 1970 the government implemented the colonial policy on commercial livestock ranching without any major policy changes. This period witnessed a very rapid expansion of the Uganda livestock industry, with the financial support from both the national government and from multi-lateral funding agencies. Uganda became self-supporting in meat and dairy products, putting an end to previous massive importation of fat slaughter cattle and milk from Tanzania and Kenya. During the same period, epidemic diseases such as rinderpest and CBPP were eradicated, while vector borne diseases such as trypanosomiasis and tick-borne diseases were controlled to a manageable level. The later part of this period, however, marked the beginning of contradictions between the post-colonial state and USAID, the major financiers of the ranching experiment. The major point of contention was an attempt to shield the ranching project from local political patronage. The radical nationalists interpreted this attempt as an infringement on the rights of the independent state.

The first five years of the Amin era witnessed a continued performance, developed during the first phase, of livestock ranching in the Ankole and Masaka schemes. In these schemes, all basic ranching infrastructures had been put in place by 1972 when Amin came to power. Two million small scale farms that had been developed owned over 95 percent of the national herd of about 5.5 million, with the remaining 5 percent on private and parastatal dairy farms and ranches. However, the ranching schemes of Lwemiyaga and Mawogola, Buruli, Singo and Bunyoro, developed after 1972, suffered from a lot of political interference in their allocation and functioning, and were not subsidised by the government in the provision of basic ranching infrastructure. By 1978 the political and economic crisis had directly affected all the ranching schemes and the commercial livestock ranching sector alike, due to critical scarcity of veterinary input supplies. From then on, the commercial livestock ranching sector began to decline considerably.

The periods 1962 to 1971 and 1972 to 1978/9 were characterised by regimes that pursued commercial livestock ranching at the expense of the traditional livestock sector. As a result, the national herd experienced a precipitous decline. The civil wars that ensued following the overthrow of the Amin government in 1979 made a situation, which was already bad, even worse.

The Obote II government between 1980 and 1985 was more pre-occupied with staying in power. There were some attempts to develop the livestock ranching sector during this period, although these efforts were over-shadowed by the civil war. The ensuing breakdown in livestock disease control infrastructure, veterinary extension services and input supplies led to the re-emergence of various livestock diseases. Diseases, which had been controlled, assumed an epidemic proportion, adversely affecting the livestock industry. By the mid 1980's, there was a complete shortage of all livestock products, with the result that the country was nearly entirely dependent on milk powder and butter oil to produce reconstituted milk for the major urban areas.

Since the colonial period, all the governments have attempted to transform livestock production. It is only the NRM government that has tailored the development of livestock production to the need to directly transform the livelihoods of traditional pastoralists. Underlying the commitment by the government to transform the traditional cattle-keeping sector has been a conviction that the ways and practices of traditional pastoralists are ecologically destructive and economically untenable. The practice of nomadism, associated with traditional pastoralism, is believed to be a leading cause of the spread of cattle related diseases. Due to the resource insecurity that it encumbers, it leads to the accumulation of herds, since there is an absence of incentive to limit the size of the herds. This creates over-stocking, which leads to overgrazing.

There have been numerous internal and external pressures on the environment in which the cattle keepers live that has been used to justify the need to transform pastoralism from being inherently nomadic. The populations of both humans and livestock have increased tremendously. Much of the grazing areas previously accessible to the cattle keepers had been alienated as a result of increased settled crop cultivation, the establishment of private ranches, military installation and farming schemes, etc. This has increasingly made the traditional mobile grazing systems of seasonally tracking resources as and when they are, available more intricate. Yet it allowed cattle keepers to use wetter and more endowed areas only for critical dry season grazing, and the drier areas during the rest of year.

The success of policy interventions designed to improve livestock production require an understanding of the underlying bottlenecks faced by the various categories of cattle keepers. The objectives of any such interventions also need to be clearly stated. For example, while the objectives of the ranch restructuring were clearly stated, there were no viable strategies put in place to deal with some of the constraints that hindered the functioning of restructured units. One of the main constraints is the availability of adequate and reliable water for livestock throughout the year; each restructured ranch should have an adequate water supply. Considering the topography, climate and runoff rates in the schemes, valley tanks and dams alone might not be able to cope with the demand. This therefore calls for the consideration of alternatives. In this regard, this study recommends the consideration of the option of sinking bore-holes to tap water reserves below the surface. But this might not provide a long-term solution for the acute water scarcity on the ranches. A more sustainable, but more costly undertaking involves or will involve the pumping of water from the rivers and lakes through pipes to a central reservoir so that a gravitational flow of water to various ranches is made possible.

Government interventions characterised by the use of force to discourage mobility and encourage sedentarisation, attempted to transform the livelihood of pastoralists in Karamoja. Today there is a demand for a high degree of organisation that accommodates flexibility to allow for quick responses. Concerns for security pose challenges and is the single determinant for pastoral livelihoods in Karamoja. A successful pastoralist today is is able to play the game of survival in such a way that security of the stocks is paramount. The preceding discussion shows that the chances of the Karimojong to maintain their adaptation to the pastoral niche depended on the effective leadership of the *alomar,* both for defence and attack. The institution of the *alomar* has developed formations that are not limited to the age-system per se. It has become more of a reconfigured system of hierarchy that has left the responsibility of security to the warlords and has left the elders in charge of cultural and ceremonial matters. We have also seen how the institution of the *alomar* has developed in response to the proliferation of firearms in the region, which necessitated vigilance and co-operation to ensure the safety of livestock through strategic defence against the competition of others.

The preceding discussion has also shown how factors external to Karamoja, especially national politics in Uganda, the war in Southern Sudan, as well as the situation in the Horn region in general, have had an impact on the lives, choices and options available to the Karimojong. The role the outside world has played in contributing to present day Karamoja is significant. It has not only led to the escalation of armament that has resulted in the insecurity plaguing the region, but it has also offered an outlet for the raided cattle encouraging continued escalation of the situation. Such factors have impinged on the social, political and economic viability of most households. The resultant institutions and forms of organisation discussed above are an adaptive response to such changes. The pastoralists systems of management are characterised by mobility and flexibility and any limitations that are imposed on this flexibility threatens the viability of their mode of production if no practical options are offered. The changing nature and character of raids have taken on a militant form that has called to question the role of traditional institutions that preached norms of social peace.

The Karamojong age-system has seen division of power of the elders who were seen as the lynchpin of authority in the past. Today the elders retain ceremonial powers, but the military decisions have been taken over by the warlords. Whereas all acts of stealing or robbery of cattle have been termed raids as long as they involve a Karimojong, the Karimojong themselves differentiate between cattle thefts and raids. When a few individuals organise to steal cattle, it is referred to as theft, but where a large organised party is involved, it is considered a raid. Whenever a few individuals were involved in a raid, the elders (even the warlords) often discounted this scale of raid as an act of banditry by 'bad boys'. It is these 'bandits' who skipped the control of not only the elders but also of the warriors, vigilantes and the state. On the other hand, the power of the warlords is recognised and does not fall under a category that is independent of the authority of the elders. Their mandate is to ensure the safety of the community under them, and to 'acquire' cattle in an organised manner. The raids that are carried out with the sanction of the elders are large and often well organised. Such raids receive the blessings of the elders. However, it is difficult to draw a fine line between thefts and raids because sometimes a small raid is a strategic move that is meant not to raise the suspicion of the victim.

Pastoralists should not be looked upon as a people with identical characteristics. Attempts should rather be made to understand the peculiarities that characterise different subsets of pastoralists. The case of Karamoja shows how adaptive pastoralists are to the various social, economic and political environments that confront them, and this calls to question the widely acclaimed assumption of pastoralists being resistant to change. I would rather pose the following question: What kind of changes do they resist, the changes that foster their adaptation, or the changes that put their survival at risk?

References

Published Sources

Ahmed, Abdel Ghaffar M. & Hassan A. Abdel Ati, 1996. *Managing Scarcity: Human Adaptations in East African Drylands.* Proceedings of a Regional Workshop held 24-26 August 1995. Addis Ababa Ethiopia. Organisation of Social Science Research in Eastern and Southern Africa.

Alchian, A. and H. Demsetz, 1973. 'The Property Rights Paradigm.' *Journal of Economic History*, Vol. 33, (1).

Ault, E.D. and L.G. Rutman, 1979. 'The Development of Individual Rights to Property in Tribal Africa.' *The Journal of Law and Economics*, Vol. 22 (1), October 1979.

Azarya, Victor, 1996. *Nomads and the state in Africa: the Political Roots of Marginality in Africa.* Leiden: African Studies Research Centre.

Baker, R., 1975. 'Development and the Pastoral Peoples of Karamoja, North-Eastern Uganda: An Example of the treatment of Symptoms.' In: Monod, T. (ed.) *Pastoralism in Tropical Africa* (pp. 187-202). London: Oxford University Press.

Barber, P.J., 1964. 'Karamoja in 1910.' *Uganda Journal*, 28, (1) 15-23.

Bardhan, P., 1993. 'Symposium on Management of Local Commons.' *Journal of Economic Perspectives*, Vol. 7 (4).

Bates, Daniel G. & Susan H. Lees (ed.), 1996. *Case Studies in Human Ecology.* New York: Plenum Press.

Baxter, P.T.W. (ed.), 1991. *When the Grass is Gone: Development Intervention in African Arid Lands.* Seminar Proceedings No. 25. Uppsala: The Scandinavian Institute of African Studies.

Baxter, P.T.W., 1975. 'Some Consequences of Sedentarization for social relationships.' In: Monod, T. (ed.), *Pastoralism in Tropical Africa* (pp. 206-225). London: Oxford University Press.

Baxter, P.T.W., 1993. The 'New Pastoralist: An Overview' (pp. 143-158). In: Markakis, J. (ed.), *Conflict and the decline of Pastoralism in the Horn of Africa.* London: Macmillan Press Ltd.

Behnke, Roy and I. Scoones, 1993. *Rethinking Range Ecology: Implications for Rangeland Management in Africa.* London: The commonwealth Secretariat, Overseas Development Institute and IIED.

Bennet, J.W., 1984. *Political Ecology and Development projects affecting pastoral peoples in East Africa.* (Research Paper No. 80). Madison Wisconsin: Land Tenure Center.

Berkes, F. (ed.), 1989. *Common Property Resources: Ecology and Community-based Sustainable Development.* London: Belhaven Press.

Brandstrom, P., Hultin, J. and J. Lindstrom, 1979. *Aspects of Agro-Pastoralism in East Africa* (Research Report No. 51). Uppsala Sweden: Scandinavian Institute of African Studies.

Bromley, D., 1989. 'The Other Land Reform.' *World Development*, Vol. 17 (6).

Bromley, D. and M. Cernea, 1989. *The Management of Common Property Natural Resources: Some Conceptual and Operational Fallacies.* World Bank Discussion Paper 57. The World Bank: Washington, DC.

Bruggeman, H., 1994. *Pastoral Women and Livestock Management: Examples from Northern Uganda and Central Chad.* IIED Issue Paper: London.

Cisterino, Mario, 1979. *Karamoja: The Human Zoo.* Wales: Centre for Development Studies Swansea.

Claudia, J.C., 1977. *Pastoralism in Crisis. The Dasanetch and the Ethiopian Lands* (Research Paper No. 180). Illinois Chicago: The University of Chicago.

Clay, W. Jason, 1984. *The Eviction of Banyaruanda: The Story Behind the Refugee Crisis in South West Uganda.* Cambridge: Cultural Survival Inc.

Cox, S.J.B., 1985. 'No Tragedy of the Commons.' *Environmental Ethics*, Vol. 7.

Dasgupta, P., 1985. 'The Environment as a Commodity.' *Oxford Review of Economic Policy*, Vol. 16, (1), spring.

Ddungu, E., 1993. *The Other Side of Land Issues in Buganda: Pastoral Crisis and the Squatter Movement in Ssembabule Sub-District.* CBR Working Paper No. 43.

Dietz, T., 1990. *Preventive and Curative Coping Mechanisms and Survival Strategies: A Summary of Concepts and An Example from the Semi-Pastoral Pokot in Kenya/Uganda.* A paper for the Colloquium on Pastoral Economies in Africa and Long-term Responses to Drought, Aberdeen University African Studies Group, 9-10 April 1990.

Dietz, T., 1993. 'The State, the Market, and the Decline of Pastoralism: Challenging some Myths, with Evidence from Western Pokot in Kenya/Uganda' (pp. 83-97). In: Markakis, J. (ed.) *Conflict and the Decline of Pastoralism in the Horn of Africa.* London: Macmillan Press Ltd.

Doornbos, M., 1993. 'Pasture and Polis: The Roots and Political Marginalization of Somali Pastoralism' (pp. 100-120). In: Markakis, J. (ed.) *Conflict and the Decline of Pastoralism in the Horn of Africa.* London: Macmillan Press Ltd.

Doornbos, Martin and Michael Lofchie, 1971. 'Ranching and Scheming: A Case Study of the Ankole Ranching Scheme.' In: Lofchie (ed.), 1971. *The State of the Nations: Constraints on Development in Independent Africa.* Berkeley: University of California Press, pp. 165-187.

Dyson-Hudson, N. and R. Dyson-Hudson, 1969, 'Subsistence Herding in Uganda.' *Scientific American,* 220, Vol. 2.

Dyson-Hudson, N. and R. Dyson-Hudson, 1970. 'The Food Production System of a Semi-Nomadic Society: The Karimojong, Uganda.' In: McLaoughlin, Peter F. M.

(ed.), *African Food Production Systems: Case and Theory*. The Johns Hopkins Press, Baltimore.
Dyson-Hudson, N., 1966. *Karimojong Politics*. Oxford: Clarendon Press.
Feder, G. and R. Noronha, 1987. 'Land Rights Systems and Agricultural Development in Sub-Saharan Africa.' *World Bank Research Observer*, No. 2.
Fratkin, E., Galvin, K.A. and E.A. Roth (eds.), 1994. *African Pastoralist Systems, An integrated Approach*. Lynne Rienner Publishers, Inc. Boulder, Colorado.
Fukui, K. and J. Markakis (eds.), 1994. *Ethnicity and Conflict in the Horn of Africa*, London: James Currey; Ohio: Ohio University Press.
Galaty, J. and P. Bonte (ed.), 1991. *Introduction. Herders, Warriors, and Traders*. Boulder, CO: Westview Press.
Galaty, J. G. and Johnson D.L. (ed.), 1990. 'Pastoral Systems in Global Perspective' (pp. 1-30). In: *The World of Pastoralism*. New York: Guilford Press, 1990.
Galaty, J.G, Aronson, D. and C.P. Salzman (eds.), 1981. *The Future of Pastoral Peoples. Proceedings of a Conference Held in Nairobi, Kenya, 4-8 August 1980*. Ottawa: IDRC.
Gartrell, B., 1988. 'Prelude to disaster: the case of Karamoja.' In: Johnson, D.H. & D.M. Anderson (eds.), 1988. *The Ecology of Survival: Case studies from Northeast African History*, Lester Crook: London.
Gilles, J.L. and J. Gefu, 1990. 'Nomads, Ranchers, and the State: The Sociocultural Aspects of Pastoralism' (pp. 99-116). In: Galaty, J. G. and D.L. Johnson (ed.), 1990. *The World of Pastoralism*. New York: Guilford Press.
Hardin, G., 1968. 'The Tragedy of the Commons.' *Science*, 162: 1243-1248.
Hardin, G., 1995. 'The Tragedy of the Commons' (pp. 38-45). In: Conca, K., Alberty M. and G.D. Dabelko (eds.), 1995. *Green Planet Blues*. Boulder CO: Westview Press.
Herskovits, M.J., 1926. 'The Cattle Complex in East Africa.' *American Anthropologist*, Vol. 28.
Hutchinson, S.E., 1996. *Nuer Dilemmas*. Berkeley: University of California Press.
IBRD/IDA, 1966. *Report on the Uganda Development Corporation on a Project to Develop Beef ranching in Uganda*. Report Number P.10. Nairobi, Kenya, December.
Jacobs, Allan, 1975. 'Maasai Pastoralism in Historical perspective.' In: Monod, T., 1975. *Pastoralism in Tropical Africa* (pp. 406-421). London: Oxford University Press.
Kanabi-Nsubuga, H.S., 1984. *Ankole-Masaka Cattle Ranching Project in southwestern*. In: Nestle, B., 1984. *Development of Animal Production Systems*. Amsterdam: Elsevier Science Publishers B.V.
Kanabi-Nsubuga, H.S., 1989. *Integration of Domestic Animals into Human Society: Inaugural Lecture*. 15 November 1989. Kampala. Department of Animal Sciences, Makerere University.
Kerven, Carol, 1992. *Customary Commerce – A Historical Re-assessment of Pastoral Livestock Marketing in Africa*. Agricultural Occasional Paper 15. London: Overseas Development Institute, (ODI).
Lamphear, John, 1976. *The traditional History of the Jie of Uganda*. Oxford: Clarendon Press.

Lane, C. (ed.), 1998. *Custodians of the Commons: Pastoral Land Tenure in East and West Africa*. London: Earthscan Publication.

Lane, C., 1990. *Barabaig Natural Resource Management: Sustainable Land uses Under Threat of Destruction*, Discussion Paper No. 12. Geneva: UNRISD.

Lane, C. and R. Moorehead, 1994. *Who Should Own the Range? New Thinking on Pastoral Resource Tenure in Dryland Africa*. London, IIED, Pastoral Land Tenure Series No. 3.

Langdale-Brown, I., Osmaston, H.A. and J.G. Wilson, 1964. *The Vegetation of Uganda and its Bearing on Land use*. Entebbe: Government Printers.

Lund, Christian and Henrik Secher Marcussen, 1994. *Access, Control and Management of Natural Resources in Sub-Saharan Africa – Methodological Considerations*. Occasional Paper No. 13. International Development Studies, Roskilde University, Denmark.

Maknun, G., 1993. 'The Decline of Afar Pastoralism' (pp. 45-60). In: Markakis, J. (ed.), 1993. *Conflict and the Decline of Pastoralism in the Horn of Africa*. London: Macmillan Press Ltd.

Mamdani, M., 1996. *Citizen and Subject*. Princeton: Princeton University Press.

Mamdani, M., Kasoma, P.M.B. and A.B. Katende, 1992. *Karamoja: Ecology and History* (Working Paper No. 22). Kampala Uganda: CBR Publications.

Manger, Leif *et al*, 1996. *Survival on Meagre Resources, Hadendowa Pastoralism in the Red Sea Hills*. Nordiska Afrikainstitutet, Uppsala.

Manger, Leif, 1995. 'Human Adaptations in East African Drylands: the Dilemma of Concepts and Approaches.' In: Ahmed, Abdel Ghaffar M. & Hassan A. Abdel Ati, 1996. *Managing Scarcity: Human Adaptations in East African Drylands*. Proceedings of a Regional Workshop held 24-26 August 1995, Addis Ababa Ethiopia. Organisation of Social Science Research in Eastern and Southern Africa.

Marcussen, Henrik (ed.), 1993. *Institutional Issues in Natural Resource Management*. Occasional Paper No. 9. Roskilde: International Development Studies, Roskilde University, Denmark.

Markakis, J., 1987. *National and Class Conflict in the Horn of Africa*. Cambridge University Press.

Markakis, John (ed.), 1993. *Conflict and the Decline of Pastoralism in the Horn of Africa*. London: Macmillan and Institute of Social Studies.

Markakis, John, 1998. *Resource Conflict in the Horn of Africa. London*: Sage Publications. Oslo: International Peace Research Institute.

McCabe, J. T., 1994. 'Mobility and Land Use Among African Pastotalists: Old Conceptual Problems and New interpretations.' In: Fratkin, E. *et al* (eds.), 1994. *African Pastoralist Systems, An integrated Approach*. Lynne Rienner Publishers, Inc. Boulder, Colorado.

McCay, Bonnie and James Acheson (eds.), 1987. *The Question of the Commons: The Culture and Ecology of Communal Resources*. Tuscon: The University of Arizona Press.

McGrane, Bernard, 1989. *Beyond Anthropology*. New York: Columbia University Press.

Museveni, Y.K., 1997. *Sowing the Mustard Seed: The Struggle for Freedom and Democracy in Uganda*. London. Macmillan.

North, D.C., 1990. *Institutions, Institutional Change and Economic Performance*. Cambridge University Press, Cambridge.

Novelli, Bruno, 1988. *Aspects of Karimojong Ethnosociology*. Italy: Verona.

Ocan C.E., 1992. *Pastoral Crisis in Northern Uganda: The Changing Significance of Cattle Raids* (Working Paper No. 21). Kampala Uganda: CBR Publications.

Okudi, B., 1992. *Causes and Effects of the 1980 Famine in Karamoja* (Working Paper No. 23). Kampala Uganda: CBR Publications.

Ostrom, E., 1990. *Governing the Commons: The Evolution of Institutions for Collective Action*. Cambridge: Cambridge University Press.

Ostrom, E., and R. Gardner, 1993. 'Coping With Asymmetries in the Commons: self-governing Irrigation Systems Can Work.' *Journal of Economic Perspectives*, Vol. 7 (4), Fall.

Pearce, D.B.E. and A. Markandya, 1990. *Sustainable Development: Economics and Environment in the Third World*. Aldershot and Vermont: Edward Edgar Publishing Limited.

Potkanski, Tomasz, 1994. *Property Concepts, Herding Patterns and Management of Natural Resources among the Ngorongoro and Salei Maasai of Tanzania*. Pastoral Land Tenure Series No. 6, IIED Drylands Programme.

Raikes, L.P., 1981. *Livestock Development and Policy in East Africa*. Uppsala: Scandinavian Institute of African Studies.

Runge, F., 1986. 'Common Property and Collective Action in Economic Development.' *World Development*, Vol. 14, (5).

Salih, Mohammed M.A., 1993. 'Pastoralists and the war in Southern Sudan: The Ngok Dinka/Humr Conflict in South Kordofan' (pp. 16-29). In: Markakis, J. (ed.), 1993. *Conflict and the decline of Pastoralism in the Horn of Africa*. London: Macmillan Press Ltd.

Salzman, C.P. with Sadala Edward, 1980. *When Nomads Settle: Processes of Sedentarisation as Adaptation and Response*. New York: Bergin Publishers.

Sanford, Stephen, 1983. *Management of Pastoral Development in the Third World*. London: Wiley, Chichester and Overseas Development Institute.

Scoones, Ian (ed.), 1995. *Living With Uncertainty: New Directions in Pastoral Development in Africa*. London: International Institute for Environment and Development.

Seabright, P., 1993. 'Managing Local Commons: Theoretical Issues in Incentive Design.' *Journal of Economic Perspectives*, Vol. 7 (4), Fall.

Swift, J., 1975. 'Pastoral Nomadism as a form of land use: The Twareg of the Adrar n Iforas.' In: Monod, T., 1975. *Pastoralism in Tropical Africa* (pp. 443-453). London: Oxford University Press.

Swift, J., 1990. *The Economics of Traditional Nomadic Pastoralism: The Twareg of the Adrar n Iforas* (Mali). Unpublished Doctoral Dissertation, University of Sussex Britain.

Swift, J., Toulmin, C. and C. Sheila, 1990. *Providing Services to Nomadic People* (Staff Working Paper No. 8). New York: UNICEF.

Thomas, Elizabeth Marshall, 1966. *Warrior Herdsmen.* London: Mwrtin Secker and Warburg Limited.

Turnbull, C.M., 1972. *The Mountain People.* New York: Simon and Schuster Inc.

Umali-Deininger, D. and C. Maguire (eds.), 1995. *Agriculture in Liberalizing Economies: Changing Roles for Government.* Proceedings of the Fourteenth Agricultural Sector Symposium. Washington, DC: The World Bank.

UNDP/FAO, 1967. *East African Livestock Survey, Regional – Kenya, Tanzania, Uganda. Volume I: Development Requirements.* FAO: Rome.

UNDP/FAO, 1967. *East African Livestock Survey, Regional – Kenya, Tanzania, Uganda. Volume II: Development Plans.* FAO: Rome.

UNSO, 1991. *Government of the Republic of Uganda: Karamoja Development Plan* (UNSO/UGA/89/X53). Kampala: The New Vision Printing and Publishing Corporation.

Wabwire, E., *The Role of Non-Governmental Organisations (NGOs) in Karamoja* (Working Paper No. 31). Kampala Uganda: CBR Publications.

Wade, R., 1987. 'The Management of Common Property Resources: Collective Action as an Alternative to Privatization of State Regulation.' *Cambridge Journal of Economics,* Vol. 11, (2), June.

Wayland, E.J. and N.V. Brasnet, 1938. *Soil Erosion and Water Supplies in Uganda.* Geographical Survey of Uganda, Memoir No. IV. Entebbe: Government Printers.

Zwanernberg, R.M.A. and A. King, 1975. *An Economic History of Kenya and Uganda: 1800-1970.* London, Basingstoke: Macmillan Press.

Unpublished Sources

Busenene, Franco, *The Management of Lake Mburo National Park Rangelands.* Kampala: Uganda National Parks/LMNP Planning Project, March 1993.

GAT-Consult and Norconsult A.B., *The Study of Water for Livestock and Domestic Use and the Related Socio-Economic and Environmental Issues in Lake Mburo National Park and Environs.* Draft Consultancy Report for the Uganda National Parks, October 1992.

GTZ/IPDP, 1997. *Monitoring and Evaluation system: A Household Survey report,* unpublished report, GTZ/IPDP, Mbarara.

Infield, M, and A. Namara, 1998. *The Influence of a Community Conservation Programme on farmers and pastoralist Communities around lake Mburo National Park: A Report of a Socio-economic Survey carried out under the Community Conservation for Uganda Wildlife Authority Project.* Kampala: African Wildlife Foundation.

Kamugisha, R. Jones and Michael Stahl (eds.), 1993. *Parks and People: Pastoralists and Wildlife – Proceedings from a Seminar on Environmental Degradation in and around Lake Mburo National Park, Uganda.* Nairobi: SIDA.

Kanabi-Nsubuga, H.S., 1977. *Livestock Development in Uganda with particular reference to the Ankole-Masaka Cattle Ranching Project.* Ph.D. thesis submitted to Makerere University (unpublished).

Kisamba-Mugerwa, W., 1995. *The Impact of Individualization on Common Grazing land Resources in Uganda.* Ph.D. Thesis, Makerere University, unpublished.

Pulkol, David, 1991. *Resettlement and Integration of pastoralists in the National Economy: The Case of the ranches Restructuring in South Western Uganda.* A paper presented at the African conference on Settlement and the Environment, Kampala, 7-11, October.

Randal, Baker, 1967. *Environmental Influences on Cattle Marketing in Karamoja.* Occasional Paper No. 5, Department of Geography, Makerere University, Kampala.

Government Reports Studied

Official Reports

Colonial Office, 1955. *The East African Royal Commission 1953-55*: Report Cmd. No. 9475. London: Her Majesty's Stationery Office.

Fintecs Consultants *et al. Meat Production Master Plan*, prepared for the Ministry of Agriculture, Animal Industry and Fisheries and the Government of the Republic of Uganda (Draft), September 1997, Kampala.

Ministry of Agriculture, Animal Industry and Fisheries, 1996. *Modernization of Agriculture in Uganda: The Way Forward, 1996-2000.* Entebbe: Ministry of Agriculture, Animal Industry and Fisheries.

Ranches Restructuring Board,1994. *Report by the Ranches Restructuring Board, Ministry of Agriculture Animal Industry and Fisheries.* Kampala (Unedited pre-draft report).

Republic of Uganda/Parliament of Uganda. *Report of the Select Committee on the Ministry of Agriculture, Animal Industry and Fisheries*, Parliament of Uganda, Parliament Buildings, Kampala, Uganda, March 1999.

Republic of Uganda/Parliament of Uganda. *Official Report of the Parliamentary Debates (Hansards), 4th Session, Second Meeting, Issue No. 14, 28th June – 23 August, 1990.* Entebbe: Uganda Printing and Publishing Corporation.

Republic of Uganda/Ministry of Agriculture, Animal Industry and Fisheries, 1997. *Budget/Policy Statement for 1997/98.* Presented by H.E. Dr. Specioza Wandira-Kazibwe, Vice President and Minister of Agriculture, Animal Industry and Fisheries. Entebbe: Uganda Printing and Publishing Corporation.

Republic of Uganda/Ministry of Agriculture, Animal Industry and Fisheries, 1998. *Budget/Policy Statement for 1998/99.* Presented by H.E. Dr. Specioza Wandira-Kazibwe, Vice President and Minister of Agriculture, Animal Industry and Fisheries. Entebbe: Uganda Printing and Publishing Corporation.

Republic of Uganda, 1988. *Report to the Government of Uganda by the Commission of Inquiry into Government Ranching Schemes.* Entebbe: Government Printers.

Republic of Uganda, 1984. *Review of the Second Beef Ranching Development Project, Final Report.* Volume 2, annex 1, Livestock Sector Review. Hunting Technical Services: Borehamwood, MacDonald and Partners: Cambridge, Consorta: Kampala.

Other Reports

Bunoti, C.W., *Progressive Report on Singo, Buruli and Bunyoro Ranching Scheme,* June 1991.

Government of Uganda, *Ranches Restructuring Board, Interim Report,* Kampala. June 1991.

Republic of Uganda, 1992. *The Ranches Restructuring Board (RRB): Second Progress Report.* Kampala: RRB Secretariat, January.

Rwamuhanda, E.R., *Progress Report of Ankole, Kabula and Masaka Ranching Scheme, Kampala,* March 1991.

Files from the Archives

Files from the Ministry of Agriculture, Animal Industry and Fisheries (MAAIF), Entebbe.

File No. C/A3, Ranches Selection Board (RSB).

File No. C.120 (a), Ranching Selection Board (RSB).

File No. C.120, opened July 1969, Ranching Advisory Committee (RAC)

File No. C.120, opened 5 December, 1963, Ranching Policy Advisory Board (RPAB), Ankole,

File SDS/51: Singo and Buruli Ranches.

File No. C.40: Singo and Buruli Ranches.

File No. C.108A Ankole/Masaka Ranching Scheme (Progress Reports).

File No. C.108/C, opened June 1970: Mawogola Ranching Scheme.

File No. C.2, Correspondences – Maruzi Ranching Scheme.

Files from the National Archives, Entebbe.

File A44/19, UPSMP No. 159/1908: Exportation of Cows to EAP, National Archives, Entebbe.

Other Files Studied

File Lan/10, Vol. 4, Land Matters, Office of the Chief Administrative Officer, Luwero.

Annual Reports Studied

Annual Report on the Veterinary Department for the year ended 31 December, 1938, Entebbe, Government Printers, 1939, National Archives, Entebbe.

Annual Report on the Veterinary Department for the year ended 31 December 1951, Entebbe, Government Printers, 1952, National Archives, Entebbe.

Newspapers

Financial Times Newspaper (now defunct)
The New Vision Newspaper
The Star Newspaper (now defunct)
The Uganda Gazette
Weekly Topic Newspaper (now defunct)
The People Newspaper

Appendices

Appendix 1: Guideline for Allocation of Ranches, 1966

1 Applicants: Ranches are to be allocated to individual Ugandans or co-operatives and Companies whose membership is made up of Ugandans.
2 Educational Background: The applicant or his bona-fide ranch manager must be able to read, write, keep accounts and generally be able to understand written or verbal instructions. Less formal education will be deemed necessary for men with experience and business ability as compared to younger and less experienced applicants.
3 Occupation: A background of successful cattle rearing and farm operation is desirable. However, proven ability in other business fields will be favourably considered, particularly co-operatives and companies with successful histories.
4 Financial Capacity: (a) 1st consideration shall be given to applicants with approximately 30,000 East African shillings or more capital available in cash and or cattle; (b) other applicants with high qualification in other categories may be considered further on basis of special loan or credit to meet financial requirements; (c) interested parties desiring to put in application, who obviously cannot qualify on financial capacity are encouraged to join a co-operative or form companies to enable them to become a partner in a successful ranch application.
5 Confidential Check of Reference: For those applicants having been considered as satisfactory under items 1-4, a confidential check of Bank and government references are made.
6 Personal Interview: After an applicant has been passed on items 1-5, a personal interview is arranged. Additional information may be secured at this time. Personal achievements, maturity, integrity, standing in local community and personal impressions are considered after this interview.
7 Pre-selection: All applicants considered to have satisfactorily met these pre-selection requirements are then passed onto the final selection Board who weigh all factors in selecting those applicants considered to have the greatest opportunity for successful ranch operation which are then documented for ranch allocation on conditional lease.
8 Approval: Applicants recommended for ranch allocation on conditional lease are reviewed by the full selection committee before final lease approval.

Appendix 2: The Composition of the Ranching Advisory Committee (RAC) Constituted on 10 July, 1969

1. The Secretary for Planning (Chairman)
2. Livestock Improvement officer (Secretary)
3. Commissioner for Land and Survey (member)
4. Commissioner for Veterinary Services (member)
5. E.M. Laki (member)
6. M.S. Sebowa (member)
7. N. Sebugwawo (member)
8. Kakiiza (member)
9. A. Kamese (member)
10. P. Oola (member)
11. Hon. J. Ekaju (member)
12. Hon. A.N. Nume (member)
13. Sezi Amot (member)
14. Representative of Uganda Development Corporation (UDC) (member)
15. Mrs. Lubega (National Assembly) (member)
16. Secretary to the Treasury (member)

Appendix 3: The Composition of the Ranches Restructuring Board (RRB)

1. Hon. David Pulkol, Deputy Minister of Minerals and water development (Chairman).
2. Mr. B. Kwarijuka (RIP), (Secretary)
3. Hon. L. Bwambale (Mrs.)
4. Hon. D. Magezi (Dr.) (RIP)
5. Hon. John Kaija
6. Hon. Latigo-Olal
7. Mr. E.R. Kirimani
8. Mr. John Mary Kasozi
9. Mr. Francis Mungoya

Appendix 4: The Presidential Committee Appointed to Assist the RRB

The President also appointed a 14-man committee to assist the Board in performing its functions, comprising of the following:

1. Hon. Elly Karuhanga
2. Hon. K. Rutemba
3. Hon. Higiro Semajege (Dr.)
4. Hon. G.W. Kasujja
5. Hon. S. Kiyingi
6. Hon. F.R. Munyiga
7. Dr. L.W. Bunoti

8 Dr. E.R. Rwamuhanda
9 Lt. Col. J. Mugume
10 Major Sonko Lutaya
11 Mr. E. Mbyesiza
12 Mr. A. Bwiragura
13 Ms. B. Musoke
14 Mr. O.A. Obel

Appendix 5: The Terms of Reference of the Ranches Restructuring Board

1 The Board may co-opt any person to assist in the performance of its functions.
2 The Board shall regulate its own procedures.
3 The functions of the Board shall be:

i To implement the resolution of the National Resistance Council of the 24th August 1990 in relation to Government allocation of ranches in Ankole, Singo, Buruli and Masindi with a view to facilitating the following:

a The revocation by the government of leases of those ranches which have not been developed by the leases in accordance with the prescribed terms and condition of allocation.

b The restructuring and sub-division of existing ranches into appropriate units; and

c The orderly and harmonious resettlement of squatters within the areas covered by the ranches.

ii For purposes of carrying out its functions, under sub-paragraph (1) of this paragraph and subject to any directions by the minister of Agriculture, Animal Industries and Fisheries, the Board shall act in accordance with the following guidelines and execute the following terms of reference:

a To undertake a thorough study of the state of each ranch before recommending whether or not government should revoke the allocation and repossess any land in which the leasee has failed to show sufficient interest and has not developed the ranch over a reasonable period or before sub-dividing the ranch;

b To implement government policy on restructuring of ranches in the government sponsored ranching schemes in Ankole, Masaka, Singo, Buruli and Masindi for the purpose of resettling the landless people with their livestock now squatting in ranches, which requires that all government sponsored ranching schemes be subdivided and categorized into smaller units of 240 hectares and 480 hectares. In exceptional circumstances, where a leasee has fulfilled most of the prescribed terms and conditions of allocation, he may be allocated 720 hectares or more, but no person shall be allocated more than 1200 hectares in any case;

c To allocate the repossessed ranches to squatters after restructuring;

d Where only part of the ranch is repossessed, to recommend to the Uganda land Commission the issuance of fresh and extension of full term title to the lessee and to existing settled squatters for their respective portions;
e To prescribe terms and conditions for the improved management of the allocated land in order to ensure sustainable livestock production and protection of the environment;
f To receive and examine any claims for compensation by the former squatters and to make necessary recommendations to the Ministry of Lands and Survey regarding such claims.
g To assist the former squatters and ranchers in forming legal bodies, e.g. co-operative societies, associations, companies or groups for the proper utilization of the allocated ranches in order to facilitate the optimal use of government services; from the restructuring of ranches and that the beneficiaries shall pay any such fees and costs as shall be determined by the government;
h To monitor the progress of ranch restructuring and from time to time evaluate the success of this policy, deeming the Minister of agriculture, Animal Industry and Fisheries informed all the time.
i To ensure that no person or persons in positions of leadership in public service, the army or any other organ of government, shall, merely by virtue of his position, acquire land repossessed as a result of the restructuring of ranches;
j To advise government on appropriate measures that the Board may consider necessary for the expeditious implementation of the restructuring of ranches and related matters;
k To examine and recommend to government the methods, costs and the means of providing infrastructure like dams, road networks, etc. that will service the restructured ranches and the land apportioned to the squatters;
l To examine and recommend to government the costs of surveying such land, the premium and ground rent payable by lessees and by squatters to who land is allocated;
m To perform any duty of function connected with of incidental to the restructuring of ranches and the resettlement of the squatters as may be directed by the Minister.

1 The Board:
i Shall commence its work as soon as possible and shall submit its final report to the Minister of Agriculture Animal Industries and Fisheries not later than eight months after the commencement of work or within such other longer period as the Minister after consultation with the President may, by writing addressed to the Chairman direct.
ii May at any time before the submission of its final report submit to the Minister any interim report on any matter within its jurisdiction as may be required or as may be determined by the Board.

Index